User-Centred Graphic Design

Mass Communications and Social Change

JORGE FRASCARA

With contributions by

BERND MEURER, JAN VAN TOORN
and
DIETMAR WINKLER,

and a literature review by
ZOE STRICKLER

UK Taylor & Francis Ltd, 1 Gunpowder Square, London EC4A 3DE
USA Taylor & Francis Inc., 1900 Frost Road, Suite 101, Bristol, PA 19007

British Library Cataloguing in Publication Data

A catalogue record for this book is available from the British Library

ISBN 0-7484 0142-3 (cased)
ISBN 0-7484-0672-7 (paperback)

Library of Congress Cataloging Publication Data, are available

Cover design by

Computerset in Times 10/12pt by Solidus (Bristol) Ltd

Printed in Great Britain by T. J. Press (Padstow) Ltd

User-Centred Graphic Design

Mass Communications and Social Change

To Tomás Maldonado,
for questioning so much, for connecting so much,
and for opening ways to a demanding, critical and responsible thinking about design

Photo: Magda Straszynka

Jorge Frascara, Professor of Art and Design at the University of Alberta, Edmonton, Canada.

Contents

Contributors ix
Acknowledgements xii
Introduction 1

1 Mapping the terrain
1.1 Visual communication design: a working profile 3
1.2 Discipline and interdiscipline 5
1.3 The audience 8
1.4 The designer 11
1.5 Design, meaning, order and freedom 28
1.6 Summary 31

2 Design methods
2.1 The quantifiable and the human dimension 33
2.2 The insufficiency of semiotics 37
2.3 Some markers in the field 39
2.4 The visualization of strategies 40
2.5 Sorting requirements 42
2.6 The question of validity in data collection *Zoe Strickler* 43

3 Targeting communications: traffic safety project report
3.1 Introduction 61
3.2 Creating a background 61
3.3 Profile of the target group 68
3.4 The focus group sessions 71
3.5 Narrowing down the target group 93
3.6 Recommendations for a communication campaign strategy 95
3.7 Future action 98
3.8 The campaign concept: focus and choices 99
3.9 Visualizing ideas 100

4 Case histories

4.1 Introduction 107
4.2 Australia's Transport Accident Commission Campaign 107
4.3 Britain's Health and Social Security forms 112
4.4 Australia's Capita Insurance Company forms 116
4.5 British Telecom: the telephone book project 117

5 Profiling the communication designer

5.1 Introduction 119
5.2 The transformation of design *Bernd Meurer* 119
5.3 Deschooling and learning in design education *Jan van Toorn* 126
5.4 Design practice and education: moving beyond the Bauhaus model *Dietmar Winkler* 129

Bibliography 137
Index 141

Contributors

Jorge Frascara is Professor of the Department of Art and Design, at the University of Alberta in Edmonton, Canada. He is a Fellow of the Society of Graphic Designers of Canada; a member of the board of the Graphic Design Education Association (GDEA); a member of the ICOGRADA Past-Presidents Forum; and a member of the editorial boards of *Design Issues* (Carnegie Mellon University/MIT), *Information Design Journal* (Milton Keynes, England), *Graphic Design Journal* (Society of Graphic Designers of Canada, Ottawa) and *Tipográfica* (Buenos Aires, Argentina). He is also a member of the Advisory Board, Department of Design, Carnegie Mellon University, Pittsburgh; chairman, ICOGRADA/education; a member of the International Institute for Information Design (IIID); and a member of the ISO TC 145, Graphic Symbols.

He was chairman of Art and Design, University of Alberta (1981–6); president of ICOGRADA (International Council of Graphic Design Associations, 1985–7); convener of the working group Visual Design Criteria for Public Information Symbols of the ISO (International Standards Organization, 1977–83); member of the National Council of the Society of Graphic Designers of Canada (1979–81); member of the Advisory Committee on Graphic Symbols of the Standards Council of Canada (1979–90); founding member, *Magenta Design Review* (Mexico, 1981); associate editor of *Icographic Magazine* (1981–4); co-chairman of the ICOGRADA-UNESCO Graphic Design for Development Conference (Nairobi, 1987); Alberta representative, Design Strategy Forum, Communications Canada (1990–1); co-organizer of the Novice Driver Education Working Conference (Edmonton, 1993); and co-chairman of Edmonton '95: Charting the Future of Graphic Design Education (Edmonton, 1995). He has organized several international design education projects, and has been a reviewer for several design education programmes.

Professor Frascara has lectured and made presentations in twenty countries, juried many design exhibitions and published several monographs and articles on design and art and design education. He is the author of *Diseño Gráfico y Comunicación* (Infinito: Argentina, 1988), and the editor of the *ISO Technical Report 7239, Design and Application of Public Information Symbols* (ISO: Geneva, 1983) and of *Graphic Design, World Views* (Kodansha: Tokyo and New York, 1990). He has conducted many sponsored research projects and produced two major research reports as principal investigator: *Research and Development of Safety Symbols* (for the Standards Council of Canada, 1986), and *Traffic Safety in Alberta* (for the Alberta Solicitor General and the Alberta

Motor Association, 1992). His professional experience includes illustration, film animation, advertising and graphic design, and he now concentrates on research and development of visual communications for safety and other social concerns.

Bernd Meurer is Professor of Design at the Fachhochschule Darmstadt (Faculty of Design), as well as founder and Head of the Laboratorium der Zivilisation – Akademie Deutscher Werkbund.

From 1960 to 1968, he was Research Architect and Assistant Professor at the Hochschule für Gestaltung Ulm. In 1965–6 he was Visiting Assistant Professor at the Research and Graduate Center of the Texas A & M University's School of Architecture in the USA. Thereafter he worked on building and heading the Holzapfel Projekt Institut (Freudenstadt) and the Projekt Co' (Ulm). Since 1973, he has been Professor of Design at the Fachhochschule Darmstadt. From 1982 to 1986, he taught at the Faculty of Humanities and Philosophy of Philipps Universität Marburg. In 1991 and 1992, he was Visiting Professor at the State University of California at San Jose. In 1994, he was Visiting Professor at Musashino University in Tokyo. He is a member of the Deutscher Werkbund, of the City Development Council of Darmstadt and of the Scientific Committee of the Istituto Europeo di Design at Cagliari, Italy.

He has published, among others, the following books: *Kritik der Alltagskultur* (A Critique of Everyday Culture, with Hartmut Vincon, 1979); *Industrielle Aesthetik* (Industrial Aesthetics, with Hartmut Vincon, 1983); *Der Rechte Winkel von Ulm* (The Right Angle of Ulm, 1987); and *Die Zukunft des Raums/The Future of Space* (1994). Some of his latest publications are: 'The birth of contemporary design', in *History of Industrial Design*, Milan, 1991; 'Modernité – l'école d'Ulm – la création réflexive', in *Design, miroir du siècle*, Paris, 1993; and 'Il traguardo', in *Laboratorium der Zivilisation*, Milan, 1993.

Zoe Strickler is Assistant Professor of Design at the Department of Art and Art History of the University of Connecticut at Storrs. She began her education at the University of Idaho, and continued it at the Minneapolis College of Art and Design where she received a BFA in 1981. In 1982, she joined the graphic design firm of Eaton & Associates Design Company, and later worked at Cohen Little Design in Minneapolis, before starting her own practice in 1987. This practice has been recognized by awards from the American Institute of Graphic Arts, the Society of Typographic Artists, *Print Magazine* and the International Association of Business Communicators. From 1990 to 1993, she studied at the University of Alberta in Edmonton, Canada, where she received the Master of Design degree in 1993 in the area of visual communications for a thesis on traffic safety communications, with an emphasis on the 18–24-year-old male problem driver. She has taught design since 1985, at the Minneapolis College of Art and Design, the University of Alberta, the University of Minnesota and the College of St Catherine in St Paul.

Jan van Toorn is Director of the Jan van Eyck Akademie in Maastricht, the Netherlands. He also teaches at the graduate programme of the Rhode Island School of Design in the USA, and has taught at several universities at home and abroad, including the Rietveld Academy in Amsterdam. He is a member of the Alliance Graphique Internationale.

He has been a freelance designer since 1967, and his work includes long associations with the Stedelijk Museum, Eindhoven; the unconventional printer Mart Spruijt; the Ministry of Cultural Affairs; the Visual Arts Centre De Beyerd, Breda; the State Department of Public Works; and the Dutch Post Office. He was the designer of the Dutch contribution, *Beyond Shelter,* to the 1976 Venice Biennale, and co-editor and designer of *Museum in Motion* (The Hague, 1979).

His work and ideas have been discussed in many publications, including the catalogue *Jan van Toorn, ontwerper*, De Beyerd, Breda, 1986; 'Typography of Jan van Toorn', by Shigeru Watano, in *Idea* **194**, Tokyo, 1986; *Spirals*, RISD, Providence, 1991, books 2 and 8; *Grafische Vormgeving in Nederland na 1945*, by Ada Lopez Cardozo and Ada Stroeve, and 'De Opleidingen', by Jeroen van den Heuvel, in *Holland in Vorm,* The Hague, 1986; 'The designer unmasked', by Gerard Forde in *Eye* **1** (2), 1992; 'Jan van Toorn', in the *Encyclopedia of Graphic Design and Designers*, by Allan and Isabella Livingstone, New York and London, 1992; 'Jan van Toorn', in *20th Century Design and Designers*, by Guy Julier, London and New York, 1993; and 'Thinking the visual: essayistic fragments on communicative action' by Jan van Toorn, in *And Justice for All ...*, Jan van Eyck Akademie series, Maastricht, 1994.

Dietmar Winkler is Professor of Design at the University of Massachusetts at Darmouth. He is a member of the Alliance Graphique Internationale, and has been an active contributor to several professional organizations. He was Director of the Institute of Design at the Illinois Institute of Technology, and Dean of the College of Visual and Performing Arts at the South Eastern Massachusetts University at North Darmouth.

He has been Design Co-Director of the Office of Publications at MIT, and Design Director at WGBH Educational Foundation Station of Boston, and at the Harvard Business Review. Since 1972, he has worked as a design consultant. His clients include the President of the United States of America's office, and the departments of Commerce; Health, Education and Welfare; Justice; Labor; and Transportation; as well as the National Endowment for the Arts, the Federal Bureau of Census, Ginn/Xerox and the New England Sports Museum. His work has been reproduced and discussed in more than twenty books and magazines. He has participated in more than forty professional juried shows, and has juried eleven other regional and national shows between 1970 and 1985.

He has conducted sponsored research on legibility, reading, retention and comprehension, and has written several articles for books and magazines, including 'Counterfeit curriculum, ersatz experience: an indictment of the present design education', *AIGA Journal of Graphic Design*, **2** (4); 'The ramification of the Bauhaus in America', *Print*, January 1991; and 'The emperor's new clothes', in *Graphic Design, World Views*, Kodansha, Tokyo & New York, 1990. He has acted as editorial adviser for Harry Abrams, Van Nostrand Reinhold and the University of California Press, and he is a member of the editorial board of the journal *Visible Language*.

Acknowledgements

I want to express my gratitude to Stella Wilson, former Director of Driver and Vehicle Programs of the Alberta Solicitor General; to the University of Alberta; to the Alberta Motor Association; and, particularly, to Robert Taylor, Director of Public and Government Affairs, for the traffic safety research support provided. My gratitude goes also goes to Karen Yurkovich, for stimulating dialogues, for her constantly inquisitive mind, and for the influence of her self-demanding attitude; to other friends, colleagues and students in the wide design community, whose sustained encouragement and example continue to motivate me; to Desmond Rochfort, for his support of my work in his role as Art and Design Department Chairman, and for promoting dialogue and enquiry in design theory; and to David Evans, who edited the manuscript. My most special thanks to Bernd Meurer, Jan van Toorn, Dietmar Winkler and to Zoe Strickler, for their willingness to contribute to this book.

I have published several of the ideas expressed in this book in articles that have appeared during the last few years in *Design Issues, Graphic Design Journal, The Humane Village Journal, Information Design Journal* and *Tipográfica*; my gratitude to these journals and their editors for allowing me to put my writings in the public space and for fostering the development of a design discourse.

The advertisements for Buick, Cadillac and Pontiac are reprinted in this book with the permission of General Motors of Canada Limited and General Motors Corporation. The advertisement for Mercury Lynx is reprinted in this book with the permission of Ford Motor Company of Canada Limited, the author being a licensee of the Ford and Mercury Lynx trademarks.

Introduction

Editorial note: quotations in this book are not essential to an understanding of the text. They are included sometimes for historical interest, to make the point that, although some ideas have been around a long time, action still lags behind; sometimes, with the intention of presenting ideas expressed in different ways, so as to allow for different styles of understanding; and sometimes for other reasons, but never pretending to strengthen an argument by demonstrating that other people agree with the main text. If the arguments do not hold, there is no use in quoting many people who err alike.

The purpose of this book is to promote visual communication design as an important resource in addressing social problems. Certainly, visual communication design by itself cannot eradicate crime, drug addiction, injury accidents or unfair discrimination, but concerted strategies, including communications, can significantly reduce the intensity of these problems. In addition to discussing communicational strategies related to social problems, this book outlines in detail some specific research and design projects that offer practical models for operation.

To accomplish its purpose, the book is divided into five sections: a discussion of visual communication design, of the ways in which it affects people, and of how its quality can be assessed; a discussion of design methods, specifically their usefulness, transferability, possibilities and limitations; a report on a project dealing with the application of visual communication design to traffic safety, aimed at providing a methodological model from which elements could be extrapolated to other attitude-changing communicational campaigns; a series of case histories, to show the feasibility and desirability of good visual communication design directed at reducing the impact of problems that affect the public; and contributions concerning the definition of the role of today's visual communication designer, which were written by other authors for this book.

The book attempts to discuss how communication design works, and the possibilities that it has to affect society positively. In this context, the book defines goals and the construction of appropriate methods. It discusses how design effectiveness could be pursued, that is, the processes that could be followed to achieve previously defined goals, and how to evaluate design effectiveness. It also considers how design practice can and should take place within a conscious ethical conception.

The book addresses the practice of visual communication design with a methodological, social and cultural emphasis. Rather than trying to simplify to offer a neat and conclusive theory, it attempts to deal with both our experience of reality and the practice of design in their complexity, because all simplification is a reduction that separates us from reality and from the consciousness of the partiality of our understanding of it. 'We recognize that there are no "facts" in science, only an infinity of possible differences (and types of difference) among which to choose to make DISTINCTIONS, and that our choice to transform or translate a particular difference into a distinction cannot not be constrained by our "hypotheses", both individual and collective' (Wilden, 1980, xxix). This book intends to make distinctions within the practice of communication design.

Visual communication design is more an interdiscipline than a discipline. As Bernd Meurer proposes, it is a problem-oriented, interdisciplinary, creative action. It cannot be defined by boundaries; they tend to be too arbitrary. It is better to see visual communication design as a crossing point of a number of disciplines which converge to create communicational objects to address human problems. It is in the intersecting of these disciplines where visual communication design operates, and it is these disciplines that give it a conceptual context and a set of tools. The knowledge of visual communication design that I am proposing (and of almost anything, I think) cannot take the form of a list, but of a multidimensional pattern, a space in constant change. This book intends to contribute to that change, outlining a large area where important communication work is needed and exploring ways to build reliable approaches for its realization.

CHAPTER ONE

Mapping the terrain

1.1 VISUAL COMMUNICATION DESIGN: A WORKING PROFILE

This book intends to assist the practice of visual communication design, proposing it as an activity directed at affecting the knowledge, the attitudes and the behaviour of people.

When visual communication design is defined this way, people assume a central role, and the visual decisions involved in the construction of messages cease to arise from presumed universal aesthetic paradigms or personal choices of the designer. Instead, they become contextualized in a field created between the people as they are now and the people as they are expected to be after having confronted the visual message. It is important to discuss how visual communications should look, but to do so within the context of what a given visual communication is, operationally speaking, supposed to do (Cross and Roy, 1975, 123). In other words, the visual dimension of visual communications should always be contextualized within the pragmatic dimension, that is, within the enactment of the communication event by the public.

Advertising design has known for a long time that if the public does not buy the advertised product, the advertising strategy has failed, regardless of the beauty of the ads or of the number of awards collected at art directors clubs' competitions. Similarly, for instance, political propaganda communications are expected to affect opinions and actions; traffic signs are intended to organize traffic flow; teaching aids are supposed to improve learning performance; and occupational safety signs are supposed to reduce injuries. The role of a visual communication does not end in its deployment, but in its effect. The motivation for its creation and the fulfilment of its purpose centre on the intention to transform an existing reality into a desirable one. This reality has to do with people, not with graphic forms.

1.1.1 Constructing effective communications

The objective of all visual communication is to effect a change in the public's knowledge, attitudes and behaviour. For such a change to happen, the communication must be detectable, discriminable, attractive, understandable and convincing. It has to be constructed on a knowledge of visual perception, human cognition and behaviour, and with consideration for the personal preferences, cognitive abilities and value systems of the audience.

Although the basic principles of perception that determine the detectability and discriminability of stimuli are, to a great extent, universal in our industrialized world, the attractiveness of something specific is more dependent upon specific subcultures. Cognitive strategies and abilities and cultural value systems vary depending on several factors, including environment, age, education, personal skills and occupation. It is evident therefore that the production of visual communications cannot ignore the specific characteristics of the public to be addressed.

There is no doubt that visual invention and dexterity are important dimensions in the creation of effective communications, but excessive attention paid to aesthetics in design, defined by the designer's values, leaves aside many genuine concerns that should be addressed in the production and evaluation of visual communications. The decontextualization of visual communication design, and the frequent concentration of design education on form construction, have led a significant number of designers to a style-conscious practice, where the qualities of the fashionable have taken centre stage and transformed the practice from a service into a field for self-expression and aesthetic enjoyment. Visual sensitivity, dexterity and sophistication, developed well beyond the ordinary person, are indispensable components in the formation of the graphic designer, not in connection with the development of a recognizable personal style but rather with the ability to construct visual messages that use, in an impacting and efficient way, the visual and cultural languages of the audience, and enrich those languages in the process. Used this way, excellence in the form of a message confers power to the communication; it results in an expansion of the viewers' visual experience; it enhances the symbolic relation between form and content; it intensifies the perceptual arousal of the viewer; it guides the act of viewing through hierarchies and sequences; it creates preciousness in the object; it generates visual pleasure; it results in a sense of awe at the intelligence and skill present in the object; and it connects the viewer with cultural values that transcend the short-term operational function of the design confronted.

1.1.2 Selling products vs. affecting attitudes

It is not indispensable to know a lot about audiences when designing communications to promote a consumer product for daily use. Given that the product does not normally differ substantially from other products, all that is needed is to effect a minor shift in the public's buying behaviour, attempting to promote one brand over another. These communications do not require a change in the public's attitudes; they ride on the back of thousands of messages that create a favourable context for the new message. The task is not easy, and billions of dollars are spent every year in commercial advertising. Generally, the difficulty with this kind of communication design lies in the lack of real differences between products, and in the need to invent them through 'image'. However difficult the task could be, the communicational efforts do not involve the generation of significant changes in the beliefs or actions of the public.

On the other hand, if the communication at stake is intended to change certain deeply ingrained attitudes in a certain group, the problem calls for a communication strategy based on a detailed knowledge of the audience and its value systems. One of the chief aims of this book is to discuss processes by which audiences can be studied and defined, and ways in which these studies can help to develop communicational strategies on which visual communications can be grounded.

To sell a product through advertising, there must be a product that, somehow or other, meets the expectations of the public. Similarly, when a communication campaign proposes a change of attitude in front of a given situation, there must be a public benefit that the public can perceive. Repressive communications do not work in the long term, and there is a need to propose an exchange, rather than a suppression, when intending to affect the attitudes and behaviours of the public. But this is not enough; although communications are essential tools to foster processes of change, they cannot generate those changes alone. It would be an impossible challenge to try to increase traffic safety through communications alone, without appropriate changes in legislation, enforcement and community participation.

If designers are to be concerned with the public's reaction to their communications, a system of evaluation must be included in the design process. The information provided by background research, and the paradigms developed to organize the design response to the problem, must always be seen as working hypotheses. Until the communication system is deployed and people have responded to it, there is no way to know with certainty that the approach taken will generate the desired results, nor is there a way to know which aspects of the strategy conceived are successful and which need adjustment. Background research and an experienced team can assist in aiming properly, but both quantitative and qualitative analyses have to take place after implementing a campaign, to provide guidance for the adjustments that communication campaigns demand, particularly when they are to be sustained for a long time.

1.1.3 Summing up

Visual communication design is concerned with the construction of visual messages meant to affect the knowledge, attitudes and behaviour of people. A communication comes to exist because someone wants to transform an existing reality into a desired one. The designer is responsible for the development of a communicational strategy, for the construction of visual instruments to implement it, and for contributing to the identification and creation of supporting measures aimed at reinforcing the likelihood of achieving the intended objectives. A careful study of the audience is indispensable, particularly when attempting to generate changes in the audience's attitudes and behaviour. Post-implementation evaluation must form part of the design strategy and serves to adjust and improve the effects of the campaign.

1.2 DISCIPLINE AND INTERDISCIPLINE

If designers are to be engaged in communications aimed at changing public attitudes towards health, safety and other social concerns, it is evident that they have to shift from having the production of the visual communications themselves as an objective to having the impact that those visual communications have on the attitudes, knowledge and behaviour of people as an objective. The production of the visual communications should be seen as a means only, as the creation of a point of interaction between current situations, desired changes and the dynamic participation of those involved.

Consumer-product advertising executives have long used interdisciplinary teams of sociologists, psychologists, technologists and marketing specialists for the development of advertising campaigns. Designers should gain from this field's existing knowledge about public response and turn it to the benefit of the public. For this to happen, designers will have to become initiators of projects and coordinators of developments. To propose the designer as a multidisciplinary coordinator, however, requires an expansion of traditional design education, so that designers could gain enough background to participate in the setting up and conducting of interdisciplinary working groups or contribute to them.

Initially, this role could be developed by large, well-established design studios, and by design educators in senior institutions. These places have the infrastructure and the resources required, both human and material, for the breaking of new ground, for the establishing of precedents, for the development of models, and for the creation of strategies. Some graduate students in several institutions have already worked on socially oriented interdisciplinary theses, related to evaluation and development of safety symbols, orientation aids for the blind, a variety of educational materials, communications for semi-literate people, and traffic safety. Graduate design education is one of the most promising possibilities for the development of new information in the field and for the creation of ambitious models of operation.

In addition to this educational approach, there is a need to develop the necessary contacts with those who can take action in the broader context of society, so as to avoid falling into a mere theoretical exercise.

> We should try to avoid, when possible, fighting programs with new programs, manifestos with new manifestos. The mere enunciation of new ideas – or supposed new ideas – is not enough. When we speak of defining our task in the age of struggle against food and housing destitution, it should be clear that the important fact is not to find an abstract (or literary) definition but an operational one: namely, a definition which is adequate to the demands of reality and which may help to guide our action successfully. Notwithstanding, an operational definition must not only clearly formulate determined objectives; it must also point out the methods for reaching them. And this is yet not all. Both objectives and methods, even when defined with the utmost empiric rigour, become transempiric, if from the start we do not question our technical, scientific or merely professional capacity to perceive those objectives and apply those methods. (Maldonado, 1989, 43 /1974, 208)

In addition to the technical, scientific and professional capacities that have been identified already in other contexts as essential components of the designer's formation, a political capacity is required. A good design proposal and the ability to implement it are by themselves not enough: such a proposal will have to be part of a larger strategy conceived by, and preferably with, those who in one way or another will be involved in the implementation of the programme.

1.2.1 Implications for design education

The education of the designer, as a problem-identifier, a problem-solver and a proactive coordinator of multidisciplinary teams concerned with the welfare of people, requires an ambitious educational programme, built to a great extent on the basis of the participation of several disciplines, whose selection and relative importance might vary from institution to institution, depending on goals and resources. The complexity of the education of this generalist, who would concentrate on the design of visual communications that address social needs and problems, would enable the designer to enter into a productive dialogue with a variety of specialists, particularly with those in sociology, psychology, anthropology, education and marketing.

Sociology is important because there is a need to contextualize the conception of the activity of the graphic designer in a frame of reference that overflows the specific bounds of the professional field and becomes grounded in the broader dimension of society. The importance of the notion of audience in the process of communication calls for a better understanding of social phenomena. The methods of enquiry used in sociology can provide graphic designers with useful instruments for the investigation of communicational problems. New methods will have to be developed, methods that possibly do not exist either in sociology or in design, and that it will only be possible to develop through interdisciplinary action. The benefit of this undertaking will go as much to sociology as to design, and, it is hoped, to society.

Psychology would be an important addition to graphic design education beyond the traditional studies in perception. Behavioural, developmental, cognitive and educational psychology would provide insights on human thinking, learning, feeling and acting at different stages in life. Experimental psychology would provide designers with valuable methods of field research in connection with specific areas of professional practice. These methods have long been used in legibility research, but their potential contribution covers a wider range of design problems.

Anthropology would provide an introduction to the understanding of culture and cultural diversity, and expose designers to value systems different from their own. Although graphic designers normally operate within their own culture, most of the time this is not entirely true, because cultural differences between people of the same geographical environment might render two groups quite different from each other. It could easily be argued that there is a subculture of designers that differs from a subculture of fishermen living in the same coastal town. Business executives, female teenagers, farmhands, steelworkers and university scientists might be considered different enough from one another to warrant a definition of their groups as representing different subcultures. To perceive, analyze and understand those differences, and to be able to use the resulting knowledge for the design of communications, an introduction to anthropology is essential.

Educational theory can provide communication designers with relevant information related to the extensive area of overlap that exists between visual communications and education. In addition to the specific design area of educational materials, many other communicational activities, up to a point, involve educational components. Such is the case, for instance, of injury prevention or family-planning campaigns, and of any other public service campaign that aims at providing the public with an extended base to make judgements about a given issue.

Marketing has been an extremely useful tool in advertising. Since the development of the concept of social marketing, an adaptation has been made of marketing theory to the conception of strategies for communications related to the common good. As a set of methods for the study of the public, marketing is an indispensable dimension for the education of communication designers.

As well as the above disciplines, the education of the designer should, of course, include relevant aspects of traditional design education, along with a good grounding in language skills.

Some readers might wonder why computing science is not among the disciplines to be included, given the popularity of the personal computer in graphic design practice. Many design schools today are dedicating substantial efforts and budgets to getting computer equipment, and this is good. But this places too much emphasis on how to produce a piece of design and too little on how to conceive it. Design is an intellectual, cultural and social activity: the technological aspect is a dependent hierarchy. Sophisticated graphic computers help now not only to produce but also to develop visual ideas. But this is not enough. A graphic computer, however sophisticated, is only a development tool, and it does not assist in the understanding of the human communication problem at hand, or in the conception of the required visual treatment.

If the computer is seen as a medium rather than as a tool, then computing science might enter the picture, because the design of computer-mediated communications requires the creation of dynamic paradigms to organize the communicational engagement between the programs and the users.

The need to form a designer in a double stream, that is, traditional design education plus social sciences and other fields, is similar to what, at a different historical time, was done at the Bauhaus. It was decided to appoint a master of form (an artist) and a master of technique (a technician or a craftsman) to educate a new professional, who was expected to synthesize visual sophistication and production knowledge.

> In the course of his training, each student of the Bauhaus had to enter a workshop of his own choice, after having completed the preliminary course. There he studied simultaneously under two masters – one a handicraft master, and the other a master of design. This idea of starting with two different groups of teachers was a necessity, because neither artists possessing sufficient technical knowledge nor craftsmen endowed with sufficient imagination for artistic problems, who could have been made the leaders of the working departments, were to be found. A new generation capable of combining both these attributes had first to be trained. In later years, the Bauhaus succeeded in placing as masters in charge of the workshops former students who were then equipped with such equivalent technical and artistic experience that the separation of the staff into masters of form and masters of technique was found to be superfluous. (Gropius, 1962, 25)

Today, some seventy years later, there is a need to extend that frame of reference, and to form a designer who would combine the model conceived at Bauhaus with a strong background in the social sciences. The increased complexity of the cultural and academic preparation of the proposed designer will certainly demand more than the convergence of two masters; there will be a need for a multidisciplinary educational experience, with a variety of possible emphases, to meet both the different areas of need and the different abilities of people.

1.3 THE AUDIENCE

Generic communications that intend to reach everybody actually reach only a few, particularly when it is intended to affect the attitudes and the behaviours of people. An emphasis on the content, such as is intended in 'Safe driving week' campaigns, without appeal to a specific audience, normally yields little result.

The mastery required by communication design – traditionally defined by a command over the language of vision – must be extended to include a knowledge of the languages, the needs, the expectations, the perceptions and the cultural values of the intended audience.

It is common today to begin any market study with the elaboration and implementation of audience segmentation criteria. Segmentation techniques vary, but it is common to talk about geographic, demographic and socioeconomic criteria. Geographic segmentation divides the market into physical areas and differentiates between urban and rural populations, in as many subcategories as necessary. Demographic segmentation considers gender, age, race, nationality, religion, marital status, and any other dimension pertaining to native or acquired qualifiers affecting the individual and the family. Socioeconomic segmentation considers income, occupation, education and, broadly speaking, social class. Other less quantitative or objective dimensions used in market segmentation attend to psychocultural dimensions, such as temperamental characteristics and social values, goals and expectations. These less quantitative dimensions consider the ways in which different groups relate, for instance, to the notions of economy, luxury, efficiency, safety, beauty, etc., and personality types such as leaders, followers, conservatives and adventurers. Finally, the particular problem to be dealt with might create its own market segmentation, such as would be the case with a specific political party, or subject areas such as ecology, poverty, AIDS, feminism or homosexuality (Chakrapani and Deal, 1992; Michman, 1991).

It is impossible to apply ready-made sets of dimensions of market research blindly to any and every communicational problem. The definition of the segmentation criteria adopted must emerge from an analysis of the problem at hand and from a definition of the critical audience that must be reached and affected.

Beyond these segmentation techniques, for a campaign to be effective, its audience must be substantial, reachable, reactive and measurable. It must be substantial, so that it justifies the effort of allocating human and economic resources to the solution of a problem through the development of a communicational strategy. Particularly when mass media or public spaces are involved, it is essential that the audience be extensive enough to justify the costs of their use, as well as the costs of conceiving a communicational strategy and giving it form. The term 'substantial' might refer to a small number of people in absolute terms, but it must be significantly large in relation to the universe of reference.

The audience must be reachable. Once it is determined that the audience is substantial, and therefore warrants the use of mass and public media to communicate with it, then the media to be used should be selected properly, so that the audience is actually reached. Certainly, it does not help to spend large sums on television announcements that are broadcast at the wrong time or through the wrong channel for a particular audience. An analysis of the specific audience's exposure to the media is essential, as well as an analysis of the kind of audience that each medium reaches.

A private school of graphic design where I was working decided one year to promote its activities on television. The campaign was expensive, but a record number of students registered that year. The problem that emerged later was that the students enrolled were not the people needed. They did not last, possibly expecting that things were going to be as simple and passive as watching television. The following year, the courses were promoted by publishing advertisements in the best newspaper in town, a newspaper of an exceptionally high quality. Fewer students came, but they knew what they were getting into and what they wanted, and attrition was negligible. Each medium creates its own public and produces a sort of market segmentation. The selection of media, therefore, should be based on knowing that a particular audience is reachable through a particular medium, and that the product, service or idea being offered is compatible with the anticipated audience. If a school director thinks that a graphic designer graduating from the school has to have specific traits, it is necessary to identify the audience that has a potential to develop those traits, and then reach it. If somebody tries to reach an audience that suffers a problem in a particularly strong way, a broad-spectrum media campaign might not be the right answer. When aiming at reaching car drivers, perhaps gas stations would be efficient channels; but when aiming at reaching public employees there is no point in broadcasting television ads during working hours.

Once an audience is reached physically, for the communication to be effective the audience has to be potentially reactive to the message, that is, a change in their knowledge, attitudes or behaviours in the desired direction should be possible. Dr Louis Francescutti, a Canadian surgeon dedicated to injury prevention, reported at a conference (Young Novice Drivers' Education, Alberta Motor Association, Edmonton, April 1993) on a rehabilitation project that he had developed and implemented for repeat offender drivers. He hosted a group of these drivers on a one-day visit to an Edmonton hospital. The group spent time in the emergency room, where they saw victims of car crashes as they were brought in. They witnessed operations. They helped feed and take care of the wounded, the paraplegic, and those in uncertain and long rehabilitation processes. At the end of the day, they all confessed to having had an extremely shocking experience, something that had changed their lives forever. Unknown to them, their driving performance was monitored for a year after this experience. It was found that nothing had changed, their driving records kept on being just as bad.

This is an extreme population profile, as far as traffic safety is concerned; however, this is a case that recurs over and over in communications that aim at affecting deeply entrenched attitudes. Generally, when dealing with issues of social interest, there is a need to identify an audience that not only is substantial in terms of causing or suffering from a problem, but also is receptive and reactive. In the case just referred to, the need for traffic safety was communicated in a strong way, but the people in the audience were just too set in their ways, and there was little that the apparently life-changing experience could do. It would not be easy to organize an effective anti-crime campaign addressed at criminals; sometimes the only thing that can be done is to address the potential victims so that the conditions make it more difficult for crime to happen. The same is true with most social problems: within the segment of the population that creates them, there will normally be a small group that will not react to communications or be persuaded in any way to change behaviour.

An audience must be measurable. If the aim of a communication campaign is to affect people, it is necessary to verify whether or not the population has indeed been affected. It is also necessary to find out to what extent it has been affected, who has been affected, which aspects of the campaign had the greatest effect and which the smallest. The main objective of this is to create feedback, so that communications can be adjusted and improved, and effectiveness and efficiency maximized.

The indices to be used to measure the success of a strategy (that is, of a campaign and related actions) should be defined at the beginning of the design process. Many times in campaigns that concern attitude change, subjects are interviewed before and after the campaign, mainly through questionnaires, and are asked about their attitudes. Normally, this is an unreliable way of finding out attitudes; a reported attitude is not necessarily an actual attitude. If a campaign is designed to reduce car crashes, the result should be fewer car crashes. That is the only real success. The durability of that success and the sustainability of the cost of implementing the campaign will depend on whether the campaign resulted only in a change of behaviours or also in a change of attitude. Only the latter will generate a positive cycle in the social group; the former will require an ongoing and constantly increasing effort to maintain it.

In broad terms, the processes of identification of a social problem, its definition, and the design of the ensuing intervention can gain clarity when looked at in epidemiological terms. In the process of identifying an epidemic, sociological statistic analyses help to define the extent of the problem, to identify vulnerable subjects and to understand the dynamics of its spread, thereby providing guidance for public health action. The study of strategies used in this field might provide socially concerned designers with methodological models that could help to increase the effectiveness of communicational programmes. This parallel drawn between design and medicine does not intend to reduce the complexities of social problems to the relative simplicity of viral epidemics, or to construct a scientific legitimation of design through a mixing of the metaphoric and the real. It intends to look for structural similarities between concerted human interventions conceived in different fields that can provide helpful models of operation to communication designers. It is indispensable, however, to pay careful attention to the peculiarities and the uniqueness of the reality at hand, before applying any imported or adapted model. Chapter 2 will deal with this aspect in detail.

1.4 THE DESIGNER

The designer gives form to the communications. This is not a simple act where options lie along a quality line that goes from good to bad. The decision-making process in graphic design is multivariate, and the information available on which to base decisions is always incomplete. Designers should strive to base their decisions as much as possible on reliable and explicable information, but the process of form-giving always calls for a leap to be taken from a series of recommendations to the creation of a form, a process that involves too many details for all to be controllable in a digital way.

The rational analysis of the problem at hand, and the verbal articulation of both the problem and the proposed action, are extremely useful elements not only for the designer's understanding of the problem but also for the client's understanding of the designer's perceptions and intentions. One of the main tasks of the designer is to create a relation of trust with the client. For this to happen, the designer must speak the language of the client (which is normally verbal, not visual), and a culturally respected language as well (which is rational, not subjective). The verbal and rational articulation of the design problem, starting with the definition of the problem itself, is therefore an essential element in the designer's task. This articulation is essential, particularly when considering designing as a problem-oriented, interdisciplinary effort.

Not all is rationality, however, because the process of form-giving calls for intuition to complement reason: not intuition understood as the naive following of gut feelings, but intuition as fast action (based on the educated use of skills acquired through extensive training) whose operational steps cannot always be in the focus of attention. This, for example, is the case with violinists, who cannot review all the nuances in the movements of their fingers while they play but have to develop the skill through sustained practice. Intuition in form-giving is the skill to make appropriate choices among infinite options available from simultaneous and interactive continuums of formal dimensions. A colour decision, for instance, automatically implies hue, tone, saturation, surface quality and a material to realize it, and it affects meaning. Thinking does not mean exclusively logical thinking, as there are other modes of thinking that become indispensable for certain tasks, particularly when exploring new areas of action. 'Bohr never trusted a purely formal or mathematical argument. "No, no," he would say, "you are not thinking; you are just being logical" ' (Frisch, 1979, 1).

The use of systematic approaches to define the content and form of a message does not necessarily lead to predictable and culturally uninteresting results. Although art in the Renaissance was essentially linked to the promotion of the ideals of the Church, the painting of the Sistine Chapel or San Francesco in Assisi was not the result of market research or the Pope's dictate; but, in addition to accomplishing the specific objective pursued, those frescoes created new dimensions for art, religion, culture and life.

Similarly, the process of form creation in design calls for something more than just following the recommendations of social and market research; it enters into the terrain of art, in the sense of being a creative act that goes beyond the mechanical implementation of information. Methodologically, the difficulty of graphic design is that a generic body of knowledge has to be applied to specific situations that relate to specific instances of human experience. 'Good planning [in architecture] I conceive to be both a science and an art. As a science, it analyzes human relationships; as an art, it coordinates human activities into a cultural synthesis' (Gropius, 1962, 142).

It should be remembered that marketing research is a 'rear-view-mirror technique', whereas design is a projective discipline. To design is to forecast, to programme, to plan future action, to create things that do not as yet exist. It is impossible therefore to ask the public about them because they cannot even imagine them. (The enemies of marketing say that, when the chairman of Sony proposed the development of the Sony Walkman, the marketing experts advised him against investing in that venture. According to them, nobody would be interested in a tape recorder that would be restricted to playback.)

The act of form-giving involves at least four distinct areas of responsibility: professional responsibility (the ability to create a message that is detectable, discriminable, attractive, understandable and convincing); ethical responsibility (the creation of a communicational engagement that recognizes the humanity of the addressees); social responsibility (the visual presentation of messages that make a positive contribution to society); and cultural responsibility (the creation of an object that enriches the cultural existence of the public, beyond the operational objectives of the design).

1.4.1 Professional responsibility

The basic performance-related concerns of a piece of visual communication design (detectability, discriminability, attractiveness, understandability and convincingness) repeats, up to a point, the sequence followed by the appearance of these concerns in the history of design.

A message must be detectable and discriminable. Other than making a message beautiful, making it detectable and discriminable have long been preoccupations of the designer. Much advertising design of this century still held these as the main concerns. But the problems of detectability and discriminability do not end in either loudness or advertising. Sophisticated perception and cognition research tests have been developed to evaluate and design highway signs, control panels, graphic symbols, teaching aids, forms and computer screen displays; nevertheless, although the collective knowledge of the profession is ready to support the design of excellent products, a great deal of this knowledge is not used for the design of most of them. This is surprising, particularly in dangerous cases such as chemical and pharmaceutical product labels, industrial plants signage, emergency instructions and securities design. In addition, the use of existing knowledge against the public is rampant in deceptive advertising. This, combined with an excessive concern with form innovation, creates a universe of visual communication products that, to a great extent, ignores this area of responsibility.

Designers often complain that graphic design is not as well respected as civil engineering because nobody can die as a consequence of bad graphic design. This is not quite true; emergency instructions and consumer product labelling are two potentially dangerous areas, and are often improperly handled. Cubana Airlines, which flies between Montreal and Havana, that is, from a French- and English-speaking city to a Spanish-speaking one, used to carry instructions for use of the life vest in Russian, because its planes were built in the former USSR. This is a situation where a communication breakdown can be deadly.

The problem of public safety does not end in the obvious areas of transportation, the workplace or open spaces. The labelling of consumer products is an area rife with legislation, but with irresponsibility as well. Legislation in many countries requires that the toxicity of substances be reported on labels; however, because there are usually no specifications of size, colour or placement, sometimes the text is set in 3-point type, the printing is of poor quality, and the colour contrast is deficient. Directions for use in 4-point type can be bothersome, although most times not harmful; but toxicity warnings in 3-point type should not be permitted. The onus is on designers to accept this as part of their professional responsibility; however, some designers frequently complain about having to include too much information on a label, feeling that this makes them unable to produce good designs. Others go even further: when legislation was introduced in Argentina compelling wine producers to print the number of millilitres contained on the labels of their bottles in typography at least 1 cm high (to help consumers distinguish easily a bottle with 950 ml from one with 1,000 ml), skilful designers rapidly produced ultralight–ultra-condensed figures that often appear more like an ornamentation of vertical lines than as numbers. Skills are available, but it is necessary to use them responsibly.

A message must be attractive. The notion of making messages attractive has also been around a long time, and it could even be argued that it is the major preoccupation of many designers today, and of many clients who decide to go to a designer rather than to a printer.

The word 'attractive' is used here, rather than 'beautiful', because it has a more clear subjective component, more directly asking the question: 'attractive to whom?'. The visualization of a message is not a self-contained process, independent from content and audience. Frequently, designs fail because of the exploration and use of visual languages foreign to the audience. Others, imitating fashionable styles, tint messages with ideologies that could be at odds with those pertaining to the intentions of their content.

A case in point concerning the intention to make messages attractive according to designers' paradigms is the work produced by the typographers of the avant-garde in the 1920s. It is true that they made a strong impact on graphic design, but the importance of this must be put in perspective by defining graphic design in a context that overflows the graphic form. The classical black and red geometric works by El Lissitsky, which have been canonized by art and design historians, were too abstract for the audience he was trying to reach. His communication-oriented pieces have not been selected by historians as anything comparable to his geometric, trend-setting work. For many art historians, aesthetic innovation seems to be the key factor in the determination of the quality of the work of an artist or designer. The constructivists had a major impact on the formal development of graphic design, but their activity and fame raised the importance of the personal aesthetic exploration to an unwarranted level in graphic design.

The excessive importance given to the avant-garde movement in graphic design history is based on the theoretical failure to recognize graphic design as something other than an art form, and to define art as something exclusively embraced by aesthetics, with insufficient consideration given to communication and sociocultural significance. Admittedly, aesthetics is an important dimension of graphic design, but it cannot be turned into a decontextualized measure of quality. The excessive importance given to aesthetics has centred the attention of designers, design educators and design historians on the formal aspects of design, that is, on the relationships of the visual elements with one another. Graphic design is, however, first and foremost human communication, and what the graphic designer does is construct a pattern, something similar to a musical score, to organize an event that becomes enacted when a viewer confronts the designed product. This tense line created between the object and the viewer triggers a response process that does or does not achieve the desired objective of the communication. Thus, the main problem that the designer faces centres on the relationship between the viewer and the visual elements and not on the relationships among those elements. Certainly, this extends the frame of reference for formal decisions, breaks their self-referentiality of forms that relate to forms, forces the designer to recognize the active participation of the viewer in the construction of the message, and establishes the importance of the viewer when discussing attractiveness.

An additional problem created by the excessive attention paid to style and aesthetics in design is that it has led to the promotion of only a few areas of graphic design, namely those that most resemble works of art, such as the poster, the book cover and the illustration. Consequently, many areas have disappeared from traditional design criticism, and it is in those very areas where communicational efficiency has been historically maintained. A discussion of attractiveness, however, should not be restricted to the notion of aesthetic pleasure, but should expand to understand the attractiveness of a clear application form that is easy to fill out, of a banknote whose value can be easily distinguished, of a manual of instructions that can be easily understood, of an endless number of things with which people interact every day and that facilitate or hamper their lives.

Often people wonder whether a visual designer should be more concerned with aesthetics or with function. This cannot be an either/or question: aesthetics is one of the functions of design. It contributes to making a design attractive to an audience, it helps to select the audience, and it contributes to the intensity of reception of the message. In addition, there is no aesthetics devoid of ideology: a given aesthetic approach expresses a world-view, promotes certain values, and influences in a subliminal way. Choosing styles haphazardly, and combining them capriciously, ignoring their implied messages, sometimes results in a neutralization of meaning, sometimes in a conflict between the forms and the intended content. Postmodern-looking advertisements for Knoll International, which sells modern furniture, are examples of a lack of understanding of the relationship that exists between the meaning of a visual style and the meaning of a product or an idea.

Aesthetics can also create another trap. When Polish poster designers, driven by an original analysis of theatre, reality and surrealism, started creating surprising integrations of men's bodies with animal heads in expressive illustrations, those images were full of intensity and content. Twenty-five years later, when those surrealistic images became an aesthetic paradigm, a recipe to repeat, they lost their power.

When the Bauhausian rationalists arrived at simplicity of design, driven by their interest in removing all traditional ornamentation from the manufacturing of objects and graphic messages, their objects had enormous power. When simplicity is sought at all cost, because it looks modern, when simplicity becomes an aesthetic norm, then the products lose their power: Herbert Bayer's alphabets with no capitals and an excessive similarity between the letters, Dieter Rams's lighters for Braun where the lighting mechanism cannot be found and where the whole object and its function are unrecognizable, glass and steel skyscrapers where all that is left is a pattern without accents, are examples where simplicity could be seen as a handicap.

The attractiveness of a message, the emotional tone of its reception by the audience and therefore the possibility of the audience remembering a message and acting accordingly are heavily influenced by aesthetics. But this has to be a content-driven aesthetics, and one that relates to the world of the audience.

A message must be understandable. The design of instructional materials for the armed forces during the Second World War, where personnel had to learn fast and well the operation of complex and dangerous equipment, had a definite impact on the increase of the importance of understandability in the visual presentation of information. Special and distance education have contributed as well. These educational and instructional developments, which took place outside the professional practice of graphic design, had a substantial and positive impact on the evolution of the notion of communicational efficiency.

Perception, seen traditionally in design as connected to form, is rather a tool for survival. We do not perceive to enjoy the environment; we perceive to understand it; and we need to understand it to survive. 'The brain is not an organ of thinking but an organ of survival' (Szent-Gyorgi, 1972–80, cited by Wilden, 1980, xxi).

The Gestalt theory is concerned with the strategies that the perceptual system uses to create coherent wholes to be interpreted. The organizing function of the perceptual system is goal-oriented, and its goal is meaning. Given the connection between perception, meaning and survival, it is easy to recognize that we have an emotional need to understand what we see. When we perceive without understanding, we experience feelings such as uneasiness, boredom, fatigue or fear, depending on the circumstances. Perception is not an act without urgency. We see to understand. We need to understand to be able to react.

Graphic communicators owe to the public the creation of understandable messages. This does not mean that everything can be made easy and clear, but to engage purposefully in the creation of obscure or ambiguous communications is abusive. Designers who engage in these games apparently like to flirt with chaos, while enjoying, at the same time, the use of an order created by others.

It is not that I am against complexity in design. The operation of complex equipment or the clauses of an insurance policy are often complex, and cannot be presented simply. 'You have explained the complex in terms of the simple – and the simple is precisely what the complex is not' (Burke, 1957, cited by Wilden, 1987, 313).

Difficult messages sometimes cannot be simplified, and in some cases an active participation of the viewer in the process of constructing the information can be an appropriate goal, to promote an active attitude or a disposition to engage in critical interpretations.

The design of instructional and educational materials, graphs, maps, tables, signage systems, graphic symbols and letters is obviously centred on the notion of understandability. Designing in these areas requires both a general knowledge of cognition and a specific knowledge of how this operates in the intended audience. Among the professional magazines on the market, no other magazine has made a better effort than the *Information Design Journal* to bring to the fore the importance of the problem and, at the same time, to provide models of operation (*Information Design Journal*, PO Box 1978, Gerrards Cross, Buckinghamshire, SL9 9BT, UK).

A message must be convincing. Every visual communication has a persuasive metacommunicational component, something that convinces the observer that it is worthwhile spending time on it, and that it is believable. The believability of a message is affected by the relationship between the values of the audience and the values expressed in the message and in the vehicle in which it appears. Clashes between values of audiences and values of messages sometimes lead not only to rejection of the content but also to lack of comprehension by an audience. A paradigmatic example is a birth-control campaign in Pakistan, where a poster showed two families: one, with two children, in which everybody appeared well-dressed and smiling, and one, with six children, showing everybody upset and a baby crying. The campaign had no visible effect. Seemingly, local people did not understand the poster. They could not fathom why the parents in the six-children family would be so upset if they had such a good number of children. Ingrained cultural preferences prevented them from understanding the poster. Had they understood it, the message might have been rejected because of value differences.

In the province of Alberta, in Canada, compulsory seat-belt legislation is observed by 87 per cent of the population. Legislation, enforcement and promotional communications show a good rate of success. But it should be remembered that this particular instance is in a country where people generally support law and order. It would be more difficult to promote seat-belt use in a tropical, developing country, where people are used to crossing the streets everywhere, where appointment times have a flexibility of about two hours, and where the relationship between people and their bodies is different, perceiving their bodies as not quite so precious as people do in the industrialized West. Different conceptions of order in general and of the value of the body dramatically affect the effect that a safety and order message can have on public behaviour.

The success of the traffic safety campaign produced by Grey Advertising in Australia in 1990 rests on a lengthy study of the elements that could make it convincing, that is, on the possible arguments that could best meet the value systems and the sensitivities of the local audience. It is not only the information contained in a message that makes people act in a given way but also a combination of factors that include the relationship between the values expressed in the message and the value system of an audience, the credibility of the source and, in some cases, such as in the example from Australia, the accompanying legislation and enforcement. Similarly, limited-time parking signs in Italy are frequently not observed; seemingly, drivers tend to believe that the ordinance will not be enforced, and that they have a right to park anywhere.

The lack of believability of sources can sometimes take dramatic turns, such as in the case of aeroplane pilots who do not believe their navigation instruments and cause disasters, because they are convinced that their own perception is more dependable than the instruments (this has been found particularly among pilots trained in an air force). In sum, the content and clarity of a message do not necessarily determine its convincing power; rather, this depends on many contextual factors that frame a message and on the believability of the source (if so-and-so says it, then it must be true, however bizarre it sounds – or, conversely, if so-and-so says it, then it must be a lie, however right it sounds).

One could recognize several levels for the concept of source of a message and its believability: the agent that signs the message, the specific vehicle in which the message appears, the class to which the vehicle belongs, the class of visual style to which the message belongs, and the specific visual quality of the message within its class. The agent that signs the message can range from an individual with a public profile (such as a fashion designer, rock star, politician or any other celebrity) to an institution as physical as a bakery or as abstract as a coalition of political parties. The specific vehicle and its class could be a given newspaper within the class of conservative newspapers, within the class of national newspapers, within the class of newspaper journalism. All these levels condition the believability and convincingness of any specific communication. The visual style of a communication also contributes to the believability of a message, and can fall under a number of labels, such as modern, postmodern, counterculture, informal, naive or downright crude. Its efficiency will be connected to whether the style used agrees with either the audience or the content of the message, or works against them or remains, somehow (not easily), inactive. There is no style that could always be the most believable. As somebody in Pentagram said, to sell fresh eggs an informal style might be better than a modernistic one, but for parachuting lessons a modernistic design would be more appropriate. In addition to the kind of style chosen for the presentation of a given content, quality also counts. Knoll International would do well using modernistic style for their communications, since they sell modern furniture, but bad quality modern-style advertising would not help.

The importance of fittingness of style to content in the convincingness of a message can be exemplified with an event that occurred during the anti-Vietnam demonstrations in 1970. On the day following a revolt that resulted in four students being killed by the police, there were demonstrations all over the USA. Students carried large banners with messages of protest. The least convincing, it seems, were the banners produced by graphic design students. Neat Helvetica, well drawn on straight lines, neutralized the power of any text. Other students, not trained in graphic design, were far more able to present moving messages, not mediated by visibly 'good' typography (related to me in conversation by Dietmar Winkler).

1.4.2 Ethical responsibility

Every situation of human communication falls within the field of ethics. That is, it can be ethical or unethical but it cannot be a-ethical. The basic tenet of ethical communication lies on the recognition of the Other – the receiver of the communication – as a subject (a person) and not as an object. By recognizing the Other as a subject, the Other is recognized as an independent, thinking person, with a specific way of understanding, evaluating and integrating experiences and information.

In ethical communications, one communicates with someone about something; one does not communicate something *to* someone.

Ethical communications presuppose a similarity between the communicating subjects, unlike military communications, where a superior communicates something to an inferior. He transmits an order that is received passively, and which is coded so as not to allow room for differing interpretations.

In ethical communications Shannon's popular terminology, borrowed from electronics and information science and defining the poles of the communication chain as transmitter and receiver, is untenable. In a universe of people engaged in communicational exchanges, it is more fitting to talk about producers and interpreters than about transmitters and receivers, terms that do not allow room for context, history, expectations, goals, values, priorities, feelings, preferences and differences of intelligence.

Shannon's terminology fosters a colonial mentality, that is, a communication model where the receivers look up at the source of the messages with passivity and reverence, and where communication becomes a unidirectional, top-down event. Particularly, this abounds in television and in mass media, where people become receivers, receptacles of communications that offer no room for interaction or reaction.

In ethical communications, the producer has to speak a language that the audience can understand. If producers really want to communicate, that is, to be understood and not just listened to, they should remember that people can only understand things that relate to things that they already understand, and that it is impossible to communicate, therefore, without using the language of the audience in both its style and content. This is why the ideal form of human communication is dialogue, where the interaction allows for exchange and adjustment, and for the building and extending of a shared terrain.

In authority-based, unilinear, top-down communications, such as is the case in the political propaganda of totalitarian governments, communications are unethical; they are conceived to be believed without being interpreted. In these cases, not only the interpreter is transformed into an object by the producer but also the producer is in turn transformed into an object by the interpreter: an object of awe or an object of hatred. Communications that offer no space for interpretation or for the construction of individual responses foster extreme responses.

The ethical dimension in visual communications is embedded in the kind of engagement created between the interpreter and the visual design. A work proposes a mode of engagement. This mode of engagement might foster a more or a less active participation of the interpreter in the construction of the message. Modes of engagement promote certain attitudes and expectations, and influence the way people relate to other people in other situations.

Many codes of conduct published by professional societies of designers mention the notion of ethics, but in many cases they only include ethical responsibilities to colleagues and clients, and seldom to the public. It is, however, necessary to relate to the public in an ethical way, that is, to recognize the public as a group of individuals, each one with a different way of understanding, feeling and acting. This is not only indispensable as a matter of principle for the creation of ethical communications but it is also important for the creation of effective ones. It is a strong asset to have the public as a partner in the promotion of changes that affect it. Imposition does not work in the long term. Although behaviours can be, to some extent, controlled through communication, pressure, fear, legislation and enforcement, if there is no partnership between the message producer and the public in relation to desirable objectives, attitudes will not change. When attitudes do not change, the need for repressive communications, legislation and enforcement constantly grows, quite likely leading finally to the collapse of the effort. Ethical communications that recognize the complexity of people and the difficulties involved in generating attitude changes from a centre of authority are the only promising approach when real changes are being sought. Although the responsibility of the designer to the client demands working with the public in partnership, simultaneously, the responsibility of the designer to the public demands the creation of a partnership between the designer and the client. Pierre Bernard puts it this way:

> If, at Grapus, we have always tried to respond to an order of a certain social type, identified by us, it is because we think that the request, wherever it comes from, generates a social act which we should be conscious of and which we should support to the fullest possible extent in our role as co-authors.
>
> This notion of co-author seems essential to me from an ethical point of view. The necessary co-operation between client and graphic designer will lead the client to share the aesthetic position (not devoid of ideology) of the designer, and the designer to accept the validity of the ideological position of his client. It is this precarious balance that allows the production to be oriented towards a cultural act, which, by definition, is always risky.
>
> I think that if this important notion does not operate in the client/graphic designer relationship then their relationship becomes a service relationship only, and we know that in our society the one who buys service is the sovereign, and the other is the submissive servant. Under these conditions, professional responsibility becomes a delusion. (Bernard, 1990, 181)

As much as this book fosters a notion based on communicating with the public and not to the public, Bernard talks about working with the client and not for the client. Seeing the client–designer relation as an opposition can lead only to frustration. It is in situations of partnership where relations become ethical, where the best talents of everyone concerned are pooled, where complex and ambitious projects become realizable, and where designers can play a role as catalysts and contributors to a constantly developing conceptual and cultural environment.

1.4.3 Social responsibility

Design, to a great extent, is market-driven, and there is a danger that the market economy will destroy the human habitat. This phenomenon will not be stoppable if industry and design are allowed to be exclusively market-driven, reactive activities. It is necessary for designers to recognize the needs of the social and physical environment within which they work and to which they contribute, and to take conscious steps to define the future direction of their profession. For this to happen, designers will have to develop new tools, engage in interdisciplinary teams, initiate projects, and generate new information and share it. This will empower more designers to engage in socially significant projects. As a result, we can expect a strengthening of the relevance, the opportunities and the perceived value of the design profession.

The improvement of access to information for the aged, for the visually handicapped and for the learning disabled; the design of educational materials for the reduction of absolute and functional literacy; the improvement of safety labels and safety signs; the improvement of the clarity of colour coding for medical and scientific research imaging processes; the improvement of typographic coding for electronic communications; the improvement of the visual presentation of forms and other governmental and administrative communications; the production of instructional and educational material aimed at the improvement of living conditions for people everywhere – these are all areas where intelligent graphic designers are desperately needed and where, at the same time, they are extremely hard to find.

This absence is partly due to the general limited understanding of the possibilities of graphic design and to the graphic designers' traditionally reactive way of operating. In most cases, the profession is seen as substantially connected to the fine arts and to the promotion of consumer goods, and this leads to a reductive perception of the role that graphic design can play in society. Seen as an artistic luxury, graphic design enters the scene of the potential users only when immediate benefits are obvious, such as is the case in the promotion of consumer goods or in political propaganda.

Expenditures on good design, however, are not such, but are investments with good returns. The cases mentioned elsewhere in this book about the traffic safety campaign of Australia and the Department of Health and Social Security of Great Britain clearly demonstrate the excellent return produced by good design. But there is more, and more that relates to the social benefits brought about by good design. According to Joyce Epstein, of the Research Institute for Consumer Affairs in Great Britain, about a million people legally entitled to public cash benefits were not receiving them before the redesign of the application forms (Epstein, 1981, 215). It seems that bad design led to social injustice by handicapping those who were entitled to the benefits but could not have access to them because of the barrier created by bad visual presentation of information. Bad design had rendered these people mentally disabled.

Disability as a function of design

Disability should not be defined by physical or mental conditions alone, but by how these conditions affect social integration and personal development. Ramps in street corners and specially designed vehicles have dramatically changed the lives of hundreds of wheelchair-bound people in many cities. As much as a bad doorknob makes a person loaded with grocery bags physically handicapped or the child-proof lid of a pharmaceutical product renders old-age users unable to open it, small type, poor layout and cumbersome presentation of text transform normal people into illiterates, keep rights away from citizens and create possibilities for crime. Indeed, pensions and benefits forms, car property registration documents and income tax forms were so cumbersome in Argentina some time ago that one needed a broker, a notary public or a lawyer to cope with their complexities, and then had to pay for those services. The problem did not end there, since people were easily cheated by these brokers, who had the power to interpret and use those cryptic bureaucratic instruments.

The designer as a problem-solver

Traditionally, designers have been defined as problem-solvers. There are two difficulties with this definition. First, the designer is not really a problem-solver, but a person who proposes a given set of actions in front of a problem out of a myriad possible actions: a mathematical problem can be solved, a design problem can only be responded to. Second, in most cases, the designer's input is framed by economic and political interests that allow little room for action about the definition of the problems confronted. Designers propose actions in response to problems; but which problems? A good graphic presentation for socially unacceptable journalistic trivia, a good storefront in an ill-conceived shopping mall, a beautiful poster for a corrupt political candidate, a good package for junk food, and an excellent logo for an environmentally irresponsible manufacturer are all examples of design understood as a problem-solving activity within a too narrow context. There is a need today to extend this context and to define the designer not only as a problem-solver but also as a problem-identifier.

The designer as a problem-identifier

Conversant as designers are with the power and the possibilities of visual communications, beyond reacting to requests for their intervention, which normally come after basic paradigms have been set, they need also to set the paradigms of their activity by identifying and defining areas where visual communication design can make a significant difference in society. Social responsibility is an active, more than a reactive concern.

There are two recognizable areas that make a difference through communication design: one relates to making life possible, and the other to making life better. Making life possible involves communications concerning health and safety. Despite the enormous expenditure every year for these kinds of communications, no measure indicates their impact. Health officials, public servants, safety engineers and others 'design' these communications, whereas well-qualified, talented and experienced designers concentrate on the creation of material for the marketing of consumer goods and services.

A permanent activation of desire is required in our market economy to keep the commercial apparatus going, and therefore it makes sense that this economy would employ many graphic designers. It is time to acknowledge, however, the existence of some hidden dimensions of this economy; for example, 51 million working days are lost in Canada because of injuries every year. Four thousand people die every year in traffic collisions in Canada, and 200 000 get injured seriously enough to require hospitalization. Five million dollars are spent on traffic injury-related health care in Canada every single day. This overloads the health services, renders them incapable of dealing with the ill, and overburdens the taxpayer.

Health care expenses keep the medical profession going, and it could be argued that the more money those professionals make, the more they pour into the market. But no wealth is produced in the process. It is imperative to look into the potential of communications related to health education and injury prevention to reduce the cost of health care. Experience shows that every dollar spent in good injury prevention strategies can save up to twenty in post-injury health care. This is important everywhere, but particularly in countries where health care is subsidized by government and paid from tax revenues that could be used for more constructive purposes. In one week in North America, 960 people die in traffic accidents, 455 are injured by handguns, 78 are killed by them, and 105 000 get injured in the workplace. On a different front, 70 000 acres of rain forest are destroyed in Africa every week, a rate that will exhaust the African rain forest in 50 years. It is unbelievable to see how societies get horrified by some things and not by others: the US press was horrified by the death of 50 000 army personnel during the 10 years of war in Vietnam, and an enormous black marble monument has been built in their memory. That same number of Americans die in car crashes every year and nobody even talks about it. There are many issues waiting for the contributions that designers can make. This is not news. Walter Gropius already stated this very clearly.

> Making a living cannot be the only aim of a young man who should want above all to realize his own creative ideas. Your problem is, therefore, how to keep the integrity of your conviction intact, how to live what you preach, and still find your pay. You may not succeed in finding a position with an architect who shares your approach in design and who could give you further guidance. Then I would suggest you take a paying job wherever you can sell your skill, but keep your interests alive by a consistent effort carried on in leisure hours. Try to build up a working team with one or two friends in your neighborhood, choose a vital topic within your community, and try to solve it, step by step, in group work. Put ceaseless effort into it, then someday you will be able to offer the public, together with your group, a well-substantiated solution for this problem, for which you have become an expert. Meanwhile, publish it, exhibit it and you may succeed in becoming an adviser to your community authorities. Create strategic centers where people are confronted with a new reality and then try to weather the inevitable stage of violent criticism until people have learned to redevelop their atrophied physical and mental capacities to make proper use of the proffered new set up. We have to discern between the vital, real needs of the people and the pattern of inertia and habit that is so often advanced as 'the will of the people.' (Gropius, 1962, 88)

All that is necessary is to dedicate a small percentage of productive time to expanding the design field toward socially oriented activities. Isotype comes to mind as an example of this kind of effort; it proves that the visual designer has possibilities to identify problems and to contribute to the development of sociocultural consciousness. It is impossible to wait until the United Nations, or NATO, or the EC, or any other 'someone else' creates these wonderful commissions. It is necessary to work toward the design of the conceptual framework within which those commissions will be grounded, and to make contacts with the appropriate agencies to initiate action.

The designer's accountability for performance

Design cannot be understood as beginning with the client call and ending with the printing of the graphic product. At both ends, there is room for design-related preoccupation and action. At one end is working on the definition of the design problem, a task traditionally left to the clients. At the other end is making sure that the operational objective of the design is achieved, that is, evaluating the performance of the intervention.

A product could be beautiful, a message could be clear, but this might not be enough. It is necessary to develop critical design performance-evaluating criteria. In the design of safety signs for a factory, for instance, the objective should not be to design the signs, but to reduce the injuries. The job will not be done when the signs go up if the workers keep on hurting themselves. These considerations make the evaluation of design quality clearer and contribute more efficiently to the reduction of problems. And these should not be defined as communicational problems only, because the objective of every communication design is, as already discussed, some kind of change that occurs in the public after the communication has taken place.

It is difficult, however, to generate the desired changes, and the implementation of campaigns concerned with complex sociocultural problems, such as handguns or birth control, has to be done carefully. The International Parenthood Federation of Africa produced graphics materials on family planning for more than 20 years in Kenya. During that period, the birth rate grew from 3 per cent per year to 4.2 per cent, to become one of the highest on earth. It seems as if the local population found the posters stimulating for some reason in a direction opposite to that intended. These complex social problems do not get solved by just doing things; things have to be done well. This requires effort, intelligence, cultural and ethical sensitivity, resources and institutional support. The design response to a social problem cannot be conceived as the production of a few posters and fliers that tell people what to do and what not to do. An example from a totally different context will show how a job can be conceived more as a communication strategy than as a drafting job.

In an intersection in New York State, close to an old people's home, the number of elderly pedestrians hit by cars reached alarming levels some time ago (National Committee for Injury Prevention and Control, 1989, 59–60). Analysis of the situation showed that:

- many cars exceeded the 30 mph speed limit;
- old people required 50 seconds to cross the boulevard, whereas the crossing time allowed by the light was 35 seconds;
- many old people could not read clearly the lit pedestrian crossing sign across the boulevard because it was too far;
- old people had problems determining the direction of the traffic; and
- many old people were not able to see the edge of the sidewalk or of the median.

The design team developed an intervention based on the following components:

- the time of the crossing light was extended according to the need;
- the distance between the pedestrians and the lit crossing sign was reduced to one-half by adding a set of lights in the median of the boulevard;
- large arrows indicating the direction of the traffic were painted on the pavement;
- curbs in sidewalks and median were painted to increase visibility;
- additional, oversized signs, posting the speed limit, were installed at various points in the boulevard;
- more police were assigned to control traffic speed;
- an extensive pedestrian education campaign for the elderly was developed in the area.

The intervention resulted in a reduction in deaths and serious injuries of 44 and 77 per cent respectively. The strategy, transferable to other situations, could be summarized as follows: identify a problem where the occurrence of a negative event is above the expected level; develop a careful analysis to determine the specific causes of those events; develop a multifrontal strategy centred around communications, but not to the exclusion of other actions; evaluate the results of the intervention.

It is obvious that in this example it would not have been enough to post a 'cross carefully' sign. A naive sign or a community-oriented poster, aimed at changing behaviours embedded in the basic values of a culture, would not work. This becomes clear when results have to be measured. The habit of accounting for the measurable effects of design interventions moves effectiveness to the foreground, and creates a clearer paradigm for design planning.

Making life better

The discussion above refers to design aimed at contributing to making life possible. On the other hand, we can think of design aimed at making life better, that is, design that would help people to discover the different dimensions of culture and pleasure. Literacy is one of the main bases of today's notion of cultural life. Much has been done about literacy, but little by graphic designers. Access to information should be regarded as an essential human right in today's society. As well, the task of the designer should not be reduced only to building invisible bridges between the information and the public. It should be wonderful to cross those bridges. 'Graphic representations of statistics should be made in such a way that they are not only correct, but also fascinating' (University of Reading, 1975, 2, citing Neurath).

Literacy means not only the ability to read, write and perform basic arithmetic operations but also the ability to understand administrative and social institutions and the rights of citizens. Reading is not an objective; it is a means to achieve better integration within today's society. It has also become clear that the acquisition of reading and writing skills is not enough, and that literacy programmes must look at long-term sustained support of the reading ability. Functional illiterates, people who learned to read and write but do not use these skills in their lives, account for roughly 25 per cent of the population in the developed Western world. Why is it that there is no written component, beyond a few words, in the lives of those people? To what extent does this render them socially handicapped in their own land, where they cannot understand an insurance policy, a political statement or a sales agreement? What are the psychological and social costs of their handicap? What can graphic designers do about this? Surely, they can do little on their own; but, surely too, they can be extremely useful for the design of visual communications aimed at assisting these people. Systematic efforts concerning literacy and the service of designers cost money, but lack of action is no cheaper. Estimates indicate that illiteracy in Canada costs $3 billion annually.

Fresh air, good food, positive interaction with others, the appreciation of art and literature, of intellectual engagement and constructive work are among the reasons it is good to be human, and supporting them is a worthwhile task, although the results of this support could not be easily quantifiable in economic terms. Perhaps not everything should be quantifiable in economic terms, but if designers want to contribute to defining and enriching the quality of life, they will have to convince those in power of the benefits of certain projects. This might involve the need to present socially meaningful proposals in political and economic terms, which are the main concerns of those in power.

If designers want to become active in the definition of their own roles, instead of allowing them to be always defined from outside, then they have to become involved in the identification of problems and in the design of the paradigms within which their work operates. To be invited after the construction of a building to create a way-finding system offers little room to deal with the way-finding problem that the building has already created. If way-finding is a concern in a public building, the concern should constrain the construction plan. If the way in which the public relates to an institution from a cognitive, operational and emotional point of view is a concern of the institution, then it is obvious that the process of communication overflows the limits of the graphic and invades the totality of the physical manifestations of the institution, beyond a conventional visual identity or a signage system. In this context, it is obvious that the designer, not as a person specialized in graphics but as an expert in visual communications, should be involved from the beginning in any process of materialization of an institution.

Normally, this is not the case today. It is common for institutions to believe that first comes the construction of the 'thing', and only after comes the problem of explaining what it is and how it works. This was the case when an aeroplane manufacturer approached Dutch designer Paul Mijksenaar to redesign the instructions for the emergency operation of the doors of an airplane. They went to him because they had received many complaints about the obscurity of the existing instructions, and they believed that Mijksenaar could do a better job. After a thorough analysis, however, he determined that there was no way of designing effective instructions, given the complexity of the mechanisms involved, and recommended a redesign of the mechanism of the opening system.

It is in the identification of problems and in the definition of paradigms where the real design problem lies. Without participation at these levels, the task of the designer becomes reduced to that of the elegant executor of someone else's conception, thereby becoming dependent on the metadesign power of the conception within which the professional participation takes place. 'We have begun to understand that, ultimately, the human environment is shaped by powers which evade our control and influence. We have been led to the more than ambiguous situation of having assumed a responsibility towards society, which in fact is exercised by others. . . . Never have we been so much needed; and never so little used' (Maldonado, 1966; 1974, 202–3; 1989, 39).

Useful action or philosophical platitudes?

The suggestion that designers should become more active in establishing the direction of their profession and its relevance to society does not imply the donation of work, but presents the need for public relations efforts aimed at demonstrating the advantages of investing in design services where there can be financial and social benefits. This will not only benefit society but also create new professional opportunities for visual communication designers.

There is a need to build a data bank of case histories, such as the ones reported in Chapter 4, to demonstrate the benefits of good communication design.

One thing has to be clear: communications alone are not going to bring about desirable social changes. If changes are desired, their achievement will depend on actions across a wide front, normally including administrative decisions, budget allocations, legislation and enforcement, that is, willingness of governments, industry and the public to take an active part. This is why designers have to become more involved in the whole process; they are more apt to know where communications can help, and when something else beyond communications will be needed to achieve the desired outcome.

1.4.4 Cultural responsibility

It is difficult to trace the boundaries between ethics, society and culture. The use of these terms has only an epistemological objective: to centre the discussion around each one of these areas while maintaining that there is nothing a-cultural or a-ethical in society.

The responsibility of the designer towards culture has normally been an interstitial concern. Pressed by clients' needs and production costs, good designers have traditionally infiltrated their work with their personal cultural and aesthetic notions. They have also tried to provide intelligent solutions, whose enjoyment went beyond the requirements of the brief and constituted a 'gift' to the public, thereby contributing to the daily enjoyment of life, and the appreciation of wit and beauty. It is important to look in more detail at the way in which this interstitial activity creates culture through the construction of models of conduct and interpersonal relations. 'Among graphic designers, there are those who conceive of their work as being based on a technical and artistic status, but with a general cultural purpose that goes beyond the simple form of an operational discourse. Therefore, for us, the goal goes beyond the message. Graphic design is a global way of intervening in the cultural debate going on in society' (Bernard, 1990, 182).

One of the major cultural responsibilities of the designer is to produce communications that actually communicate something. There is no cultural existence without communication, but the amount of utterances in today's environment creates the illusion that we are surrounded by 'the information age' where what surrounds us are generally signs without information. Advertising and political speeches are good examples of this, and journalism contributes a great deal: the commercial pressure to publish a newspaper every day has resulted in enormous amounts of redundance and irrelevance. It seems that there is a need to recover the capacity of language and images to actually communicate information, and of people to distinguish between information and noise.

At the centre of the notion of information is the notion of relevance. Relevance keeps communications alive, prompts the development of language, and creates new expressive needs that call for new visual solutions, which in turn stir the cultural environment.

> Gaston Bachelard in his book *La terre et les rêveries de la volonté* writes: 'It seems that there are already areas where literature reveals itself as an explosion of language. According to chemistry an explosion happens when the probability for ramification is greater than the probability for termination.' In final analysis, it is the explosive states – and therefore the states of intense animation provoked by the appearing of new images – that can favour communication in life. Small talk moves in the opposite sense: it reduces the probability for ramification and increases that for termination. Its scope is the perpetuation of the 'unconscious archetypes.' (Maldonado, 1974, 31)

New experiences, coupled with the desire to communicate them, foster the development of language and information. Information 'is a difference which makes a difference' (Bateson, 1972, 453). Without new information to be communicated, language becomes stale.

In 1988, the Warsaw Poster Biennale organized a one-day symposium called 'The Crisis of the Polish Poster'. When repression was in place, Solidarity had been pushed underground, corruption was rampant, supplies were scarce, and the whole system was bursting through the seams, all that the posters dealt with – and had dealt with for almost 40 years – was related to traditional theatre plays and other performing arts, including circus. In addition to this thematic restriction, posters were not to be seen outdoors in the cities. Printed in runs of between 100 and 200 copies, they were restricted to theatre lobbies, cultural events and poster exhibitions and competitions. The poster had been recontextualized, transformed into the latest addition to art for art's sake, and had lost all relevance – and possibilities – in relation to the urgent issues that moved the lives of people in Poland. Yes, the poster was in crisis, because the main preoccupation of the poster artists was to remain faithful to their personal style, submitting the viewers to the repetitive experience of recognizing the recognizable list of authors (the authors belonged to a stable list; there was no opening for new people). From the surprising form-creation explosion created by the Polish posters of the 1950s and early 1960s, which had contributed to creating culture, 30 years of repetition had reduced the information carried by every new poster to practically zero, with no visual language development to contribute to the intensification of the intellectual or aesthetic experience. They had become a series of redundant objects, standing for the cultural stagnation of their time.

Important design work, instead, contributes to save lives, and to make them functionally better, but it also adds value to life, and feeds that conscious cultural existence that rises above the biological. Design is based on projective thinking; it is a teleological, future-oriented, constructive action, developed in pursuit of a realizable utopia.

The culture of things

All objects with which we surround ourselves are a language beyond language, an extension of ourselves, a visualization of the invisible, a self-portrait, a way of introducing ourselves to others, and an essential dimension of humanity.

No animal constructs objects and surrounds itself with them to the same extent with the purpose to communicate. It could be argued that one buys a watch to know the time, a suit to cover one's body, and a car to move from A to B with freedom, but it is obvious that functionality, in the restricted sense of the word, refers only to a small part of the functions of our objects and of the reasons that we choose them. In addition to meeting their specific functions, we choose our objects to help us communicate with others, to show aspects of ourselves. And this is not all: the objects that we choose to represent ourself, in part, also build it. 'There is no doubt that here the contained and the container – the human condition and the human environment – are the result of the same dialectic process, of the same process of mutual conditioning and formation' (Maldonado, 1992, 27).

Our clothing and our furniture condition our acting. Our thinking and our attitudes are affected by the newspapers, the magazines and the books we read, by the people with whom we interact and by the places we visit. There are objects that elevate us and objects that degrade us, objects that help us to be what we want to be and objects that hamper our development.

In the court of Knossos, the throne of the king was raised only some 15 cm above the seats of the councillors. In ancient Egypt, the seat of the pharaoh was raised above the heads of his subjects. The two situations expressed and constructed the hierarchical relationship among the members of government in the two societies.

A Louis XV chair, a Marcel Breuer chair and a wicker chair invite us to adopt different attitudes in front of everything. The flimsy construction of most buildings in North America presents a world without past or future, whereas the stone buildings of Europe make one feel part of a tradition that overflows the limits of a person. The government building interiors of the Soviet Union in Moscow provide an interesting example of an opposition between purported beliefs and real feelings: eighteenth-century despotic monarchy architectural design was used to frame the 'government of the people', an environment of grandeur that would make it difficult to remember the ideas of Marx, let alone implement them – an environment, as well, that expressed very clearly the actual relation that existed between the government and the people.

It was the powerful language of things that, for different motivations, moved Pericles to build the Parthenon, the French to build the Eiffel Tower, and the Americans the Empire

State building. It is the powerful language of things that brings some people together and pulls others apart; it is a language that does not lie, because it is normally constructed on the basis of preferences for form, without intention to communicate. All designers, as builders of the daily environment, have to learn to understand it, and to speak it with appropriateness, proportion and measure, so that, perhaps, they could contribute to the building of a harmonic culture of objects that would slowly invade life itself. 'Production is the most effective medium of our time, a medium that can be used as a vehicle of stupidity or of culture' (Dino Gavina, cited by Vercelloni, 1987, 178).

It could be argued that the impact of design on our daily life, on the evolution of our habits, perceptions, preferences and attitudes, is one of the strongest we confront. Only mass media could be compared to it. Our world is populated by designed things, from manholes to cities, whereas cows and forests have become abstract concepts. The power of design lies in its pervasive subliminal presence. Unlike art, which exists framed in independent objects, the objects of design are placed everywhere, generally ouside the focus of attention. Even though there can be beautiful computer keyboards, excellent typefaces and magnificent dining-room furniture, we concentrate on our typing, reading and eating. It is for this reason that design is so powerful; it sets the tone of our experience of things, and conditions our interpersonal relations.

The nature of the function of design normally requires its invisibility. In most cases, the better the design, the more invisible it is. One uses daily objects without seeing or experiencing consciously their design until using something that is poorly designed (Alexander, 1979, 22). Small typography in warnings and directions, objects that do not work or incomprehensible maps show the absence of good design, reduce efficiency and upset people. Good design allows people to achieve their daily goals smoothly, as if there were no mediation.

Every day, from the time that the alarm clock goes off until waking up the following morning, all activities are enabled or hampered by design, be this graphic, product or environmental. Human life happens in a world conceptually promoted and sanctioned by mass media and built by design.

It is difficult to assess the cultural impact of visual communication design *vis-à-vis* other forms of communication, such as literature, art or journalism; but there is no question that advertising stereotypes promote certain social values that, however desirable or undesirable, condition the development of many aspects of culture and the connotations of many social roles. Models of ideal appearance, desirable objects and, in sum, most cultural values had been sanctioned in the past by the ruling classes with the help of art and their physical presence in the small communities of their time. Today, this is done by political and commercial interests through mass media. To make a point, art directors often tend to promote stereotypes, with a view to fulfilling the appropriate fantasies of the targeted market segment. These stereotypes, however, can prove to be a social liability when evaluated within broader contexts beyond the specific objectives of an advertisement. A clear example is an ad for an airline, where the picture of a 6′6″ cowboy, spread out in a first-class seat of an aeroplane, suggests that this particular airline offers a lot of leg room. The picture appears convincing but promotes an attitude of carelessness about those who surround the model. This becomes clear when thinking about being not the model but the person sitting next to him. Would this character not be a nuisance also in a bus, at a concert, at a gas station, anywhere? The promotion of an airline on the basis of the leg room it provides, done in this manner, also promotes a social liability. Examples such as this abound.

On the one hand, the mass media promote certain types and roles, raising their currency and value, whereas, on the other, they omit others, somehow marginalizing certain sectors. As much as North American advertising models are normally young, slender, beautiful and have perfect teeth, illustrations in socialist countries before the collapse of the USSR were just as narrow in their dictionary of types, showing construction workers, factory and industry workers in general, technologists, scientists, soldiers, and occasionally nurses and teachers. The media, however, normally forgot the bus drivers, the office clerks, the waiters, the street sweepers, the letter carriers, and many others. It is tempting to believe that postal services and restaurants were so bad in those countries because people in those sectors felt left out of the picture. The media led them to feel that they did not count and had no value in the construction of their country. Normally, mass communications promote a limited set of types in any society, defining the central and the marginal.

Advertising has often been criticized for defining roles and patterns that influence the way in which people conduct their lives. It is impossible to blame the whole field in a sweeping generalization; but, in daily practice, graphic designers working in advertising face the problematic use of cultural stereotypes. Communication is based on the sharing of codes, and therefore stereotypes are useful as a means to ensure understanding. But the audience's understanding of messages should be achieved without reinforcing models of questionable value. The problem deserves attention, and designers who are now equipped by education and by practice to meet the needs of clients should become more aware of the needs of society, and be better trained to meet them.

Images stand for values, and should be seen as propositions, particularly when they are presented by the mass media, a context of authority. According to Wittgenstein, '. . . in every proposition there lies an image-model (Bild-Abbild), and in every image-model, a proposition. Or better: in every proposition an image-model *is shown*, and in every image-model a proposition *is seen*. "The image" proposes Wittgenstein, is a model of reality (2.12) . . . The image presents the situation in a logical space, the being and the not being of states of things (2.11) . . . The proposition is an image of reality (4.01)' (Maldonado, 1993, 20, citing Wittgenstein, 1961, 15 and 37). The notion of linking models and propositions makes models acquire the double meaning of representing and proposing. This results in a double responsibility for the designer: on the one hand, the model of reality should respond to the existing reality for it to be intelligible; on the other, its existence as a proposition will affect the existing reality and reinforce its possibilities for development in a given direction, and should therefore be consciously handled.

1.5 DESIGN, MEANING, ORDER AND FREEDOM

Every perception, as already stated, involves a search for meaning. Every search for meaning requires an ordering process, and every ordering process requires a design hypothesis. This design hypothesis operates as an attempt to impose a structure of relations and hierarchies over a cluster of stimuli to create meaning.

> The observation is familiar: many phenomena perceived initially as complex (almost unintelligible or non properly representable) seem to become all of a sudden understandable as soon as the model makers 'change codes' to describe them, or to decipher the code through which they read them: the ellipse, the electron and the wave function seem to be concepts invented by Man to represent in a simple way phenomena perceived as complex such as force, energy and power. As soon as we describe them by means of that purely conceptual new code (or language), it seems possible to see as intelligible, even simple, that phenomenon that yesterday was inextricably complex. (Le Moigne, 1992, 89)

Perception is a meaning-seeking organizational task. Faced with incomprehensible chaos, different people at different ages and with different beliefs and abilities react differently. Everyone tries to understand. To understand is, first, to interpret signs and to invent connections. 'To find similarities between situations despite differences which may separate them' and 'to make sense out of ambiguous or contradictory messages' are recognized by Douglas Hofstädter as essential indicators of intelligence (Hofstädter, 1980, 26).

Second, to understand is to try out hypotheses and see if they work. One of the main tasks of the visual communication designer is to facilitate the ordering processes that perception and understanding involve. Perception is an intelligent act.

To understand involves a learning process, and every learning process involves a secondary learning (technically called 'deuterolearning'), that is, a development of the ability to learn other things similar to the one learned. If one learns by heart a senseless series of letters, not only does one learn that series, but one also learns to learn senseless series of letters. The more one exercises the ability to understand new things, the more able one becomes to understand new things, and consequently the more confident and independent one feels.

To understand involves transforming chaos into a system of meaning. For this, it is necessary to be able to interpret the signs of an environment and to create connections. Eskimos, tuaregs and jibaros know well the meaning of the small details of their respective environments. Snow, sand and jungle for the citizens of industrialized countries are only that: three things. Eskimos, however, have an extensive number of words to define the many different and significant states of the snow, as much as the tuaregs have many words for the colour of sand. The jibaros know an endless number of Amazonian plants and their properties, transforming the word 'jungle' in a useless simplification.

Similarly, the word 'city' is also a simplification. Hundreds of realities form the complex concept of 'city', and every day we discover in them still more and more realities. Although a jibaro learns everything that they should know about their environment before they turn 18, at the professional end of the industrialized West, people keep on learning all their lives, even if they never move away from their native city. Their context is not physical, it is conceptual. Life in the city today is basically life in a communications environment. This environment constantly challenges people to transform noise into information, to ignore redundance, and to separate the relevant from the irrelevant.

> We don't know how to articulate what we really need. We don't know how to describe the performance we need from a person or a product to solve the problem. We don't know, in other words, how to make constructive demands for an improved, better working environment.
>
> So, as I said, the question really is: what don't we have so that we can ask the right questions? And the answer is obvious: information. (Information in its highest form is both an art and an entertainment.)
>
> . . . What this story says is that out of the first Commission meeting came the proclamation: *Public Information Must be Public*. . . . It meant that the city of Could-Be was a society because of its citizens and that it was therefore responsible to the aforesaid citizens. It meant that all information about Could-Be concerning its past, present and future, was the property of its citizens. That both the government and the people of Could-Be had a responsibility to the city and to each other to make all information about the city accessible, and readily accessible. That is to say, not only was public information to be made physically available, it was to be made understandable as well. (Wurman, 1976, 16–17)

In our city environment, communication designers are builders of landscapes, not only expressing their culture but also shaping it. In the chaos of stimuli that constantly overloads us, and that constitutes the raw material of culture in the city, the designer could be the organizer who transforms that chaos into information.

1.5.1 The aesthetics of order

To the eyes of the nineteenth-century art lover, raised in the neoclassicist academy, the Baroque was disorderly and chaotic, a mere decadence of Renaissance art. Wölfflin argued in 1915 that the Baroque was not the decadence of the Renaissance, but that rather it was a style based on different principles, and had a different kind of complexity (Wölfflin, 1915). It was this difference that had rendered the style as apparently complex and unstructured, and it took Wölfflin an extensive effort to explain the 'code' through which Baroque art should be formally understood and perceived as orderly. Possibly, Wölfflin based his thesis on Worringer's notion of 'will for style' (Worringer, 1911). He proposed that the task at hand was to understand the organizational principles of the Baroque style itself, arguing that different intentions generate different styles, and that a problem was created when attempting a qualitative evaluation of the Baroque using Renaissance paradigms. Worringer developed his thesis in defence of the Gothic style by recognizing it not as a primitive Renaissance but as a style derived from a Germanic 'will for style', defining the concept of style as an organizing system based on a specific way of creating, selecting and combining visual elements, according to a specific cultural conception (Worringer, 1920).

Similarly, the designer of today sometimes has to build communications that involve a high degree of complexity and a move away from the usual. Certain messages and certain goals require new kinds of visual presentation, and it is essential that these new contents, however much they could need a new form, do not become obscure or chaotic to the eyes of the intended audience, who would see as too complex anything for which they do not have a code. Forms have to be appropriate, but they also have to be, to a sufficient extent, transparent.

Although Wölfflin and Worringer worked on the relativity of the aesthetic judgement and on the definition of different systems of visual organization used in different periods, what is common to the Gothic, the Renaissance, the Baroque, or any other visual style, including youth counterculture, is the existence of a system of order. This has always existed, not only in the shaping of an artistic style but also in the totality of the constructions – physical or conceptual – of a culture. 'Man is a singular creature; he has a series of gifts that make him unique among the animals, so that, different from them, he is not a figure in the landscape – he is the shaper of the landscape' (Bronowsky, 1973, 19).

The disposition of humanity to engage in the building of its landscape is the clear expression of a need to organize it according to a particular conception: simple and physical, as it can be among the Masai in Kenya, where order in daily life means the building of a circular barrier of thorny branches that surrounds the compound and keeps the lions away from people and cattle; complex and abstract as the stock market; or technological and exact as a NASA base; or hypercomplex as the city of London.

In our complex Western reality, the interaction among people and with things created by people is, to a great extent, governed by time. Time is wealth today, and its coordination is a necessary condition for the development of intense and efficient human communication. All social activities, at work and leisure, are governed by time. In this environment, coordination, order and agreement become indispensable. Precision in the information, punctuality in meeting deadlines, and speed in the actions require clear communications and are based on a respect for the time of others. It is the notion of the value of time that creates the need for efficiency, and it is to a great extent the assigning a high value to efficiency that supports the existence of design as we understand it today. Design, among other things, helps people to act efficiently.

In our highly complex reality, design and order are essential conditions for communication, work and freedom. Thanks to planning and order, it is possible to understand the underground trains in London, the telephone services in an Oslo hotel, and the income tax form in Canada. Lack of order creates dependence, as in the Argentine bureaucracy examples discussed earlier, where complexity leads to the need for mediators and makes room for crime. Much energy can be lost in disorganized societies just trying to cope with daily errands.

The order present in the daily environment penetrates the life of a place. The ideal society is organized and transparent. These two principles reinforce and facilitate each other. It is easy to make an organized society transparent, and it is easy to keep a transparent society organized. Opacity breeds corruption and chaos. The communication designer has a meaningful role to play: in the task to facilitate access to information, the designer contributes to making reality more transparent and promotes among people the habit to be informed. Informed people can understand better the complexities of society and escape the simplifications that lead to fanaticism and extreme positions. People who understand the systemic nature of society, and the ways in which things work, are more able to exercise their freedom and to better integrate themselves in society as productive members. The designer, as information provider, can become the promoter of social integration, fairness and freedom.

1.6 SUMMARY

A responsible designer responds with conscious efforts to the requirements of the client, within the context of the needs of society. Responsible designers become active in the definition of their own role, and of the paradigms within which they operate. Beyond the construction of the visual elements that constitute a communicational campaign, the designer should take a step further and participate in the conception of the realizable utopia to be pursued, and in the development of the communicational strategy to be applied. In this way, the visual construction of the messages will be tailored properly to the communicational needs, and the potential contribution of visual communications to society will actually take place. But this requires a lot of learning, details and patience. 'God is in the details', it is said that Mies van der Rohe once declared, and the truth is that the details are many, possibly constituting the maximum challenge.

Touching social problems and changing people's attitudes require more than good intentions, from both designers and clients. There are enough market-driven designers to keep the economy going, but there is a great need for talented communicators in the social marketing field, as much as there is a need to demonstrate to governments and the private sector how much benefit there is to be collected from intelligent communications in this field, even financially. There is also a need to urge good designers to promote the potential value of communications in this front and to make work in this front profitable, not only to governments and the public but also to themselves.

Five points summarize the basic ideas discussed above:

1. Visual communication design is centred on human behaviour and not on visual forms.
2. Visual communication design applied to solve social problems can learn from

epidemiology, marketing and medical therapy: as much as it is necessary to define the disease and to know the patient to administer a therapy in medicine, in socially oriented visual communications it is necessary to define the problem and to know the audience to create an effective communicational strategy and its related actions.

3. The evaluation of the outcome of a communicational campaign must be an integral part of the design plan, and it requires achievable objectives and a substantial, reachable, reactive and measurable audience.
4. A thorough knowledge of the problem and of its socioeconomic impact is necessary to secure the resources required for the support of appropriate action.
5. Visual communication design is an activity that has an impact on the public sphere, and as such requires a professional responsibility that overflows the technical. A technically excellent but ethically and socially irresponsible designer is a social, cultural and ecological hazard.

This is not an unrealistic, utopian position. It would be unrealistic to keep on exclusively responding to market needs, reactively allowing the market to define the designer's role. This would keep graphic design exclusively at the service of short-term commercial interests. It makes sense in a market economy for design to contribute to its activation; but it is necessary to direct good design to other, more hidden dimensions of the economy and the urgent needs of people.

The public should be made more aware of their basic rights and obligations. This awareness is now affected by the excessive noise created by the media, and by the lack of interest of most governments in long-term efforts, particularly when these have no consequence on immediate elections. It is evident that visual communication design has a role to play, and that the core of the professional preoccupations cannot be formed by the discussion between modernism and postmodernism, the obsession with computers or the best tricks seen in the last annual. Visual communication design has to strengthen its concern for what really matters: life, death, pain, happiness and the welfare of people.

CHAPTER TWO

Design Methods

To design the research method and to design the design method are tasks of a higher order than to design the actual communications. Methods create frames, paradigms within which design decisions take place. Their implementation is active and reflective, feeding back onto them.

2.1 THE QUANTIFIABLE AND THE HUMAN DIMENSION

The attempt to develop a methodological model in this book is based on the belief that visual communication design should develop more sophisticated methods towards an increasingly efficient and effective practice. This is not an attempt to adopt pseudo-scientific, mathematical or engineering methods, or to bring back the method fever of the 1950s and 1960s, which resulted in a situation where the intention to quantify, and to consider only that which could be quantified, left out important design problems. Today, we believe that it is necessary to confront systematically as much quantifiable variables as non-quantifiable variables in design planning. Christopher Alexander indicated this need already in 1964: 'The importance of these non-quantifiable variables is sometimes lost in the effort to be "scientific". A variable which exhibits continuous variation is easier to manipulate mathematically, and therefore seems more suitable for a scientific treatment. But although it is certainly true that the use of performance standards makes it less necessary for a designer to rely on personal experience, it also happens that the kind of mathematical optimization which quantifiable variables make possible is largely irrelevant to the design problem' (Alexander, 1979, 98–9). It is interesting to note that the work of some pioneers of the design methods movement does not promote methods as valid by themselves. The same Alexander writes: 'I reject the whole idea of design methods as a subject of study, since I think it is absurd to separate the study of designing from the practice of design' (*ibid.*, vi).

Precise methods for calculating the thickness of a cantilevered roof or the radius required by a curve in a superhighway can be devised, including all kinds of dependent variables, but mathematical formulas are not useful when human responses are at stake. Among some road safety engineers, there is interest in what they call 'the shifting target'. When a high-collision intersection is identified, they tend to introduce a number of physical changes into it to solve the problem, such as adding stop signs, turning lanes or traffic lights. These modifications improve safety in the intersection, but in a few days a new high-collision intersection usually develops one or two blocks away. The same problem exists in connection with curves in the highway. When too many drivers miss the curve and run off the road – because they drive at 130 km/h in a 100 km/h highway – the curve gets rebuilt and adjusted to make room for those 130 km/h drivers. These drivers, in turn, when they see a safer curve, take it at 160 km/h.

It is easy to estimate through precise design methods how to design a curve in a highway for cars to take it at a given speed; but, when human behaviour factors come into play, no partial approach can help. One of the failures of traffic safety understood exclusively from a road engineering point of view is to assume that people will behave lawfully. Indeed, any problem understood from one exclusive point of view and any solution attempted on the same basis will most likely be ineffective. In visual communication, the problems are not problems of visual communication, but of people. This assertion extends the frame of reference and permits a better planning of possible action, keeping in mind that the behaviours of people respond to complex and varying factors that no linear method can predict and no inflexible strategy can address. When discussing the differences between visual communication design methods and engineering, linguistics or scientific methods, one has to address the distinction between the complex and the complicated: the circuits of a supercomputer are hypercomplicated, but not complex; however, where human experience and behaviour are concerned, one certainly faces the complex. According to Le Moigne, complexity appears when one 'considers familiar problems, such as the rapport North–South (or East–West) . . . or the affective, productive or cognitive relations between two human beings engaged, for instance, in the same task' (Le Moigne, 1992, 88).

2.1.1 Precision and usability

The interest in watertight methods affects not only graphic design but also many other fields, and it becomes a negative factor when the interest in the perfection of the method separates researchers from reality. It is a bit like game theory: 'Knowledge of game theory does not make anyone a better card player, businessman or military strategist, because game theory is not primarily concerned with disclosing the optimum strategy for any particular conflict situation. It is concerned with the logic of conflict, that is, with the theory of strategy' (Rapoport, 1962, 108).

For methods enthusiasts, the object of study sometimes becomes a secondary aspect, because of the difficulties that it presents in not submitting to standard filters and universal measures, resisting the possibility to build theories with a high degree of precision. This often creates a distance between the theoretical discourse, which becomes concentrated on the definition of categories and terminologies, and the practice.

There is no problem in engaging in discussions concerning terminology, as long as the referential function of language grounds them. Precision, and general mastery of language, are important dimensions of any planning activity. Verbal formulation helps the projective thinking of the designer; however, this thinking deals both with the representation and the transformation of reality, and therefore it requires theories that not only discuss the internal structure of design but also its operational interaction with the reality that it affects. Theories and methods are only means, and means in design need to be complemented by a good perception of reality and a creative imagination to affect it according to established aims. The effort that we need to make today is to bridge theory and practice, so that theory does not remain self-referential, and practice moves beyond intuition. Methods form that bridge.

Visual communication design methods should assist the practice. They should therefore be minimal, that is, no greater than what is necessary to direct productive action. They must be flexible, that is, adaptable to the requirements of each case. Excessive and rigid methods reduce the visible scope of reality and filter what remains visible. Abraham Moles (1964) suggests that all methods are basically aleatory and that their success is never guaranteed. He argues that methods are not recipes that lead infallibly to a result and that there is no such thing as an inventing machine. If methods are too highly structured, they would turn into recipes and would lose applicability in proportion as they gain precision. This book discusses a methodological model, but the model is not to be copied but examined, adopted and adapted, when some of its components appear to be useful for projects concerning other segments of reality.

Nigel Cross and Robin Roy produced in 1975 an extremely useful book about design methods, centred around the gathering and handling of information. Most of the fifteen methods they discuss are quite instrumental, and closer to the practice than most other sets proposed in the literature. One of the main advantages of the methods discussed by Cross and Roy is that they do not constrain the activity of the designer, but promote the possibility for breaking one's own habits of thought and enquiry, by adopting different strategies for gathering and analyzing information (Cross and Roy, 1975, 5–7).

2.1.2 Fitness of method to problem

Every method used should be constantly under critical review. While the method serves to investigate reality and build actions to affect it, it has to be at all times in the focus of attention of the designer, never to be adopted as a framing device without framing it in a broader and critical context. An analysis of the problem, the context and the objectives of the design action will help to select the most appropriate method. Even the most appropriate one, however, will need to be modified according to specific characteristics of the task at hand.

> In the origin, the word method means path. Here, we have to accept to walk without a path, to make the path as we walk. It's as Machado said: 'Caminante no hay camino, se hace camino al andar' (*Walker there's no path, one makes the path as one walks*). The method cannot be developed but in the research: it cannot come to light and be formulated but after, at the moment in which the arrival becomes a new departing point, now furnished with a method. (Morin, 1992, 29)

It is important to be aware of the newness of the situation that one confronts, and to be ready to invent ways to confront it, without being deluded by the impression that previous experience or acquired knowledge can be extrapolated in a ready-made fashion. 'The brain uses ready-made subroutines to solve problems. It acts using previously developed responses created in front of what it presumes to be similar situations to others faced in the past' (Margolis, 1987, 27). The tendency of our brain to use ready-made subroutines is efficient for the rather inconsequential chores of daily life. The design of mass media campaigns to affect people's attitudes and behaviours is a different and complex thing, and it requires an aware and resourceful mind that constantly observes itself and its own strategies.

An openness to understanding the specific characteristics of a problem makes it possible for a designer to develop new ways of building information and strategies, and to construct new 'paths' that help to achieve stated objectives. It is important to devise a method, but it should stem from the problem itself. 'To arrive at the point that you don't know, you must take the road that you don't know' (St John of the Cross).

Methods help to investigate something that we do not understand well, but that we understand enough so as to choose the appropriate method of investigation. There is an interesting story that can be quoted here to illustrate the notion of appropriateness (of method to subject). It comes from Lichtenberg, an eighteenth-century thinker:

> Lichtenberg dreams. In the dream, a 'supernatural' old man, whom we recognize effortlessly as the figure of God, entrusts him with a sphere, and indicates to him a laboratory where he may analyze it. Lichtenberg very carefully manipulates the sphere, dries it, rubs it to test its electrical properties, and then, finding nothing special, analyzes it chemically and finds some one hundred simple common elements. Disillusioned, he gives the inventory to the old man who informs him that the analyzed sphere was the Earth, and describes to him the catastrophes caused by his 'hasty' manipulations: the seas evaporated at his blowing, the Andes are still attached to his handkerchief, etc ... Lichtenberg requests another chance, and the old man entrusts him with another object, inside a bag, for him to analyze chemically. Lichtenberg opens the bag and falls on his knees, defeated: it is a book. (Stengers, 1992, 80–1, citing Lichtenberg, 1844–53, vol. 6, 48–60)

The discussion of methods of enquiry and methods of construction aims at developing an awareness of the need to adapt the kind of strategy to be used to the kind of problem that one is confronting. As the Lichtenberg story shows, not knowing what the sphere was, he was unable to understand how to study it, and he did not approach it with pertinent methods. Knowing what a book was, he realized that a chemical analysis would not yield any information relevant to the existence of the book as a book. Since there is no way of confronting a problem without assumptions, it becomes imperative to develop appropriate ones and to consider the extent to which the instruments of enquiry chosen would or would not allow one to obtain relevant information, particularly when confronting something not already known.

The repeatability of an observation in science was interpreted in the past as indicating that the phenomenon studied was as described. It is now clear that all that we can be sure of is that the perceptions of the observers are repeatable. The structure of the phenomenon, however, is another thing, and it is easy to be deluded into believing that one understands something, just on the basis of agreement with other people's perceptions of the same thing.

Often, what we see as necessary is only habitual. Our cognitive system easily slips into seeing two frequently sequential things as connected by a cause–effect relationship.

The discussion of methods, therefore, is brought here to promote consciousness about the difficulty involved in understanding what we do not yet understand, to stress the importance of the connection between method and problem, and to highlight the role of the observer in the construction of the observed. 'Every piece of knowledge, whatever that could be, presupposes a knowing mind whose possibilities and whose limits are those of the human brain, and whose logic, linguistic and informational substratum comes from a given culture, and therefore from a specific society' (Morin, 1992, 113; see also Selvini Palazzoli, 1989, 82–3).

Methods exist to help in aiming the direction of action. There are methods related to the structuring of the design process, and methods for each one of the steps of that process. But any step taken should be taken as a working hypothesis, something to be tried, attempted and tested within the context in which one is operating. It is important to act systematically, and to make sure that every step taken makes sense and can be justified. The confirmation of the usefulness of each step must be sought through empirical evaluation, that is, through a comparison between the outcomes pursued and the outcomes obtained, rather than between the actions taken and the actions intended.

2.2 THE INSUFFICIENCY OF SEMIOTICS

Semiotics has been studied with interest by many designers since at least the 1950s in an attempt to devise a method to organize systematically the different aspects involved in the communication process. The discipline, however, was developed not by designers, that is, not by people engaged in the construction of communications, but by scholars who studied the phenomena of communications from outside. More than a strategy for the construction of messages, semiotics was developed as a tool for their analysis, and normally did not concern itself with the specifics of concrete audiences. Concrete audiences are not easy to reduce to simple structures of relation and subordination, and a discipline that intends to be a science might find specific audiences too difficult to codify; however, communication among humans is varied, difficult, complex and seldom predictable. Semiotics, on the other hand, stems from linguistics, and from the linguistics that deals more with the Saussurian 'langue' (the abstract structure of language) than with the 'parole' (the verbal communication between people), tending to be an analysis of language, but not an analysis of actual communications (Saussure, 1915).

Semiotics relates to cognition, but human behaviour, including cognitive behaviour, operates on a complex base that far exceeds a purely cognitive discourse. Design methodology and semiotics have gained prestige through 'naming the parts', that is, developing a terminology and a set of categories. Names have a magical power. If people can name something, they believe they understand it (some even believe that they possess it). This is what has led to the development of endless, preferably Greek or Latin, lists of names of parts of communicational phenomena, substituting denominations for explanations. It is easier to name the force of gravity than to describe it, let alone explain it.

2.2.1 To inform and to persuade

A purely cognitive analysis might suffice to begin work in a given area of visual communication design, such as the design of symbols for public information; however, a difference has to be recognized between an information symbol such as 'waiting room' and a regulatory symbol such as 'no smoking'. The first one has to be understood, and this is difficult, but the second one has to be understood and acted upon. This clearly overflows the cognitive dimension and expands the function of the sign to the edge of its possibilities. This is the problem of many traffic signs, and their lack of effectiveness, which sometimes prompts city officials to add signs that read 'Obey the signs', with unknown results. A more original approach was taken in Manhattan, where at some bus stops drivers can see a sign that reads: 'DON'T EVEN THINK ABOUT PARKING HERE'. It seems that the more colloquial and personal tone of the speech works as a better reminder of the possible presence of a police officer. The notion of authority is emphasized by the use of a text composed in capitals (people speak in lower case and scream in capitals, according to Jock Kinneir, the father of British highway signs).

Signs alone are insufficient to enforce regulations and effectively issue warnings; they have to be part of a comprehensive system of communication, including legislation, education and enforcement, obviously overflowing the field of semiotics. To a greater or lesser extent, all communications touch emotions and overflow the purely cognitive.

2.2.2 Semiotics and rhetoric

Whereas semiotics started as an analytical/descriptive endeavour, rhetoric began as a construction-oriented one. Over time, however, an excessive list of figures have continued to maintain a medieval dimension in the discipline, rendering it more prone to serve as a dissection instrument of analysis than as an aid in message production. Nevertheless, it cannot be denied that the main rhetorical categories can help the visual communicator to understand the different ways in which the structure of a message can be designed, and can also help to avoid the structural repetition that one is prone to fall into because of personal thinking habits (see Ehses, 1986, 1988).

The operational insufficiency that affects both semiotics and rhetoric is that they do not help the decision-making process beyond the display of possibilities. This is where the pragmatics of communication have to reach over for sociology, psychology and marketing, in addition to other relevant fields of knowledge, to develop the information required for the construction of effective communications. Rhetoric and semiotics fall short of this need, having always been centred on the analysis of the production of language, without dealing with its interpretation, thereby becoming too general when actual and specific responses are expected from actual and specific people, in relation to actual and specific messages that address actual and specific problems.

2.2.3 The specificity of the design problem

That concrete, individual and diverse people are at the centre of any communication problem demands a flexible position from the communication designer. Communication theory or safety engineering might work with generic postulates, but communication design has to deal with the concrete, the individual, the diverse and, up to a point and however difficult, with the uncertain and the unpredictable.

> The signs of a system are always symbols, i.e., institutionally sanctioned signs. The symbols have significance only for the members of the same 'meaning community' (J.H. Leavitt, 1951). In other words: communication is only possible when the 'stock' of signs at the disposal of the various individuals overlap (W. Meyer-Eppler, 1959). But this ideal situation, of complete congruity of the stock of signs within the overlapping field, exists only in a few artificial languages. In natural languages, on the other hand, this total congruity is out of the question. Lexicologists and semanticists know that every natural meaning community is characterized by discontinuity and heterogeneity. To overstate it: no one can quite understand the other (F. Paulham, 1929). Nevertheless, the vague and ambiguous character of natural language does not always produce disadvantages in human communication (R. Carnap, 1937). On the contrary; it is sometimes an advantage, too, for natural speech can only fulfil its function with the aid of open, free sign complexes existing in a state of constant transformation and extension (W. Mays, 1956). (Maldonado, 1959, 70; 1989, 210)

The insistence on this kind of argument does not contradict one of the basic aims of this book, that is, to stand as a defence of methodical design planning, defining the methods as working tools, and discussing the limitations that affect the transferability of methods from one situation to another. Methods are not transferable in their entirety, because, being tactical things, their quality is highly connected to their details, and their details to the details of the problem at hand. The knowledge transferability that can be discussed is the ability to approach a situation in such a way that one can, taking cognizance of the situation, define and develop the best method of approach.

2.3 SOME MARKERS IN THE FIELD

The function of design methods is mapping the field of action. Methods deal with the establishment of markers and pointers, with the description of sequences and processes, with the provision of options for action, and with the identification of factors, connections and interactions, both for the collection of information and for its use in the development of design projects. As an example of methodical sequence, the following list shows 12 essential steps that were followed for the development of the traffic safety project reported in Chapter 3. Although the headings are listed in sequential order, some actions partly overlap.

1. Identification of a problem that requires action and that can be reduced using visual communications.
2. Description and definition of the dimension of the problem. Description of who and what is affected by it, and to what extent. Human, social and financial costs. Analysis of statistical information. (This information is required before approaching competent authorities and relevant agencies that could help to develop the necessary research, and could eventually finance an intervention.)
3. Identification and definition of the causes of the problem. Human behaviours and other factors.
4. Identification and definition of a critical population that is highly affected by the problem and that forms the critical target population of the future campaign. Recognition of this population as substantial, reachable, reactive and measurable. Qualitative and quantitative information gathering. (Although the identification of the target group can be done through analysis of quantitative information – normally statistics – it is important to gather qualitative information once this is defined. Qualitative information is of a different logical type than quantitative information. The issues raised during the qualitative information-gathering process frame the design of strategies aimed at collecting quantitative information, normally in the form of questionnaires. The former identifies the existence of certain phenomena; the latter measures their frequency.)
5. Definition of operational, measurable and achievable goals.
6. Definition of strategies to achieve those goals, centring on communications, but not excluding other actions and other possible partners. General definition of style, content and channels of communication.
7. Construction of prototype messages, based on the visual and verbal culture of the target audience and constrained by the aforementioned definitions. This is the intelligent leap in the void: all the data collected to this point now has to be used to create imaginative, arresting and appropriate messages aimed at affecting people in the intended direction.
8. Testing of prototype messages by way of presenting them for opinions and reactions by sample group of target population.

9. Implementation of the campaign in a controllable, small-scale environment.
10. Further to analysis of the results of this implementation, and to appropriate adjustments, deployment of the campaign in the large-scale environment for which it was conceived. Coordination of multifrontal/multi-agency action.
11. Measurement of effects. Evaluation of the totality of the campaign and of each one of its components. Analysis and feedback.
12. Sustained continuation of the campaign with regular controls, adjustments and changes.

2.4 THE VISUALIZATION OF STRATEGIES

Step 7 of the preceding list is crucial to the conception of a communications campaign. It is where data have to be turned into images, where recommendations have to be acted on, and where a synthetic action has to merge different levels of analysis into the construction of a set of concrete objects.

> Splitting up a complex problem means hierarchizing it; the various groups of variables are thus weighted according to their relative importance. It will be apparent at once that personal judgements and prejudices inevitably creep into the design process at this point. The process of dividing up a problem can be visually represented in the form of a graph, or more specifically a 'tree' consisting of elements (= variables) and connecting lines (= reciprocal relations between the variables). At the top of such a tree stands the problem in an undifferentiated and thus insoluble form. As the ramifications increase in a downward direction, the subproblems are arranged at various levels. Analysing a problem in this way does represent an important step forward but it stops short of the product form, which is to say that the product has not yet been designed. In essence the form is contained in the 'tree'; it must therefore be decoded from the diagram and converted into an object. This process of conversion – the actual design work – has hitherto formed the arcanum of all design methodologies. Suffice it to say – without attempting any premature explanation of the fact – that so far no design methodology, not even in its most sophisticated form, e.g. that of Ch. Alexander, has proposed a conversion from an analytical diagram to a form. It is at this point therefore, that future efforts to inject methodology into the design process must begin. (Bonsiepe, 1967, 152)

Still today, some 30 years after the methodological efforts of the Ulm school, the transformation of strategies into visual realities is a weak point as far as design methods are concerned.

One of the major problems to be faced is the insistence with which existing visual paradigms attempt to creep into any project. Elsewhere in this book, the problem created by choosing an inappropriate style for a message is mentioned, along with the danger of creating a conflict between the intended meaning of a message and the actual meaning of the visual style. It is easy for a designer to be invaded by models of what could be considered 'good design', or by the visual environment and the segment of mass media to which the designer is exposed. Visual structures are perceived so unconsciously that one can immediately recognize the datedness of a 20-year-old advertisement, although being unable to describe the conditions that date it and oblivious to the specific peculiarities of today's graphics.

The move from the strategical to the tactical, and from there to the visual, poses a number of problems derived from the need to move from the general to the particular. In the traffic safety project discussed in Chapter 3, for instance, the strategy recommends bringing the value of safety above the value of thrill. But how to do it? It might be wise to use an admired reference group to 'sell' safety, or one might resort to the notion of the indirect social benefits of safe driving. But images are concrete, and they imply the choice of a particular instance of social benefit for a particular group. The recommendations established by a communicational strategy can guide the designer, up to a point, concerning the image content of a message. But the particularities of the image have to be created as an effort of interpretation. One can talk about 'an admired reference group', but which group exactly? How many people? How many males and females? What ages? How are they dressed? What are they doing? Where are they? Is it sunny, midday, evening? Are they inside an old building, a car shop, or are they in the great outdoors, or in an exotic place, or around the corner? Is the scene shot at eye level, low, high, with a wide-angle lens? How is the light? And the style of everything from the shoes to the graphics? Visualization is adamantly concrete and has to be programmed like a feature film production, carefully and thoroughly, for it to be believable and effective. This is the reason that it is so important to develop prototypes and test them with a sample of the target group. The high number of detailed decisions demanded by the actual visual production demonstrates the large margin of uncertainty left by the recommendations of a strategy plan, however extensive it could be. Given this, a first approach to visual production should always be seen as a working hypothesis.

Two main areas of decision can be distinguished in the process of visualization: image content and visual presentation.

The image content refers to the objects, people, animals, plants and environments appearing in the picture as well as the stories represented or suggested. The visual presentation involves four levels of analysis: the composition of the objects, and their place in the frame; the light; the point of view; and the graphic style of the whole.

The composition of the objects in the frame establishes hierarchies and a visual sequence (it guides the viewer from point to point on the surface according to a plan), and constructs systems of relationship among the objects represented (establishes subordinations, oppositions, levels of importance, pertinence and relevance). The light can be harsh or soft, natural or artificial, chromatic or achromatic, intense or dim, and the tonal range can be extended or restricted, and high, low or medium. The point of view has to be chosen from among different possible heights and distances, and can involve different photographic lens angles. The style can vary across the possibilities shown by the different existing products of the mass media, such as teen magazines, counterculture, sports, news, business, etc. Table 2.1 summarizes the variables involved in visual presentation.

Decisions about the visual structure of the messages will have to respond simultaneously to what is intended to be achieved and to the visual environments of the target population.

2.4.1 Texts and images

The process of visualization involves the systematic planning of the image, the structure of the verbal components, and the relationship between both codes. Following Roland Barthes's taxonomy, one can define the relation between the image and the text as being either 'anchoring' or 'relay' (Barthes, 1985, 28–30). In 'anchoring', the text focuses on and emphasizes a meaning already visible in the image. It 'anchors' it, holding down one out of the myriad possible meanings of an image. This kind of relationship contributes to the production of simple, clear and direct messages. This, however, is not necessarily an

Table 2.1 Visual variables involved in the creation of images

Image content	*Staging and props* (consider production as a feature film): time and place setting *Objects:* precious, casual, technical, etc. *People:* age, gender, class, education, temperament, dress, etc. *Environments:* geographic, social, professional, etc. *Story:* the scene as a point or as a sequence in an imaginary story	
Visual presentation	*Composition*	Positioning in the frame Hierarchies Relationship among image components
	Light	Harsh or soft Natural or artificial Colour temperature Extended or restricted chromatic range Extended or restricted tonal range High, medium or low tonal dominance
	Style	Adoption of historic or contemporary formal systems Balance, dynamism, fantasy General appearance of the whole
	Point of view	High, low, medium Far, close Lens angle

exercise in the obvious or the redundant, because images have many possible meanings and, in this situation, the text intensely focuses the viewer's attention on one of them, creating an explicit hierarchy that the image does not necessarily have by itself. In the 'relay' situation, both text and image contribute to the building of a message, which is incomplete in either of them. In this case, there appears to be a need for a more interpretive attitude on the part of the viewer to understand the message, which is incomprehensible based on the image alone.

2.5 SORTING REQUIREMENTS

As said before, one of the least developed areas of design methods is actually the transformation of requirements into forms, be these architectonic, industrial or communicational. All that information gathering provides is isolated bits, issues to pay attention to. The task of the researcher/designer is to find the patterns that connect (Bateson, 1972, 131–4). To do this, once the issues to be considered are identified, it is useful to write all the issues on individual pieces of paper and move them, creating different groupings and relationships. In this way, one can try a great variety of possibilities of organization, until one finds an operationally satisfactory one. It is important not to be afraid of contradictions. It is possible that two different and apparently opposed kinds of message are necessary in a campaign. One might need, for instance, to appeal to youths' need for solitude and relaxation, but to their love for action as well. Finding requirements of this kind should not prompt one to a choice. Often, apparent contradictions are just expressing the need for balance and complement. At the time of understanding the issues that emerge, and of structuring possible ideas and articulations, nothing should be discarded. Sometimes it is in the apparent impossibility to connect disparate things that new possibilities hide, and discovering them might change one's perspective and create new approaches to the construction of communications.

The objective is not to find the true connection between units; the objective is to find an operational organization of the various units that would render them usable for the construction of visual messages. For this purpose, one has to identify similarities, create groupings, determine hierarchies across the groups and within the groups, establish priorities, and create main headings and lines of argument. Once this is done, one has to write down a series of scripts, to develop each line of argument in several different ways, so as to address them to the different subgroups of people who have been identified within the target audience. Out of these, a few argument lines have to be selected, to serve as a basis for 'marketing' the ideas. These ideas have to be thought of as products, and media budgets have to be allocated in amounts similar to those necessary for a new product launch.

2.6 THE QUESTION OF VALIDITY IN DATA COLLECTION

Zoe Strickler

2.6.1 Data-gathering methodologies in the social sciences

Data collection in the social sciences falls into two theoretical fields defined as either 'quantitative' or 'qualitative' research. An individual researcher will usually start from one position or the other, but the decision to work from either position strongly affects the nature of the information gathered, the ways in which it is collected, and the uses to which it can be put.

Quantitative research methods are those that have as their primary objective the measurement of social phenomena and the expression of those phenomena in specific mathematical values. Qualitative methodologies are designed to describe social phenomena and express them as a set of observations. Whereas quantitative research draws its theoretical values and models from the 'pure' or natural sciences, qualitative research has its origin in the 'soft' sciences, specifically anthropology. A Gallup poll that uses sophisticated sampling techniques to express public opinion in percentages is a widely understood – and possibly misused – example of quantitative methodology. The methods of cultural observation used by Margaret Mead in her studies of Samoan culture are qualitative. Raw data collected by either quantitative or qualitative methods must be ordered and analyzed as a set of findings if the information is to have any meaning in a broader context.

A major structural difference between qualitative and quantitative methodologies is the approach taken to defining 'categories' for study (or specific, identifiable relationships between objects and phenomena). Quantitative methods follow precedent in the natural science disciplines by defining research categories as precisely as possible in advance of undertaking an actual procedure. The process involves constructing hypotheses, or 'expectations', which are then 'proved' or 'disproved' by the procedure. A primary goal of quantitative methods is to control the variables being studied to the greatest degree possible to achieve high levels of accuracy in the measurements. It is assumed at the outset of a quantitative procedure that categories for study will remain unchanged during implementation (McCracken, 1988, 16).

Qualitative methodology assumes that categories for study will be identified during the procedure itself. Procedures are designed to yield branching results and multiple hypotheses, and to expose issues and interpretations that could not have been predicted before the investigation. Qualitative researchers assume that categories will change over the course of the study with the acquisition of new, unanticipated information. The qualitative researcher, therefore, looks for patterns of relationships among many categories, rather than isolating and measuring individual phenomenal relationships. Quantitative methods are considered more effective for measuring the existence and prevalence of behaviours, whereas qualitative methods reveal the motivations behind the behaviours.

In *The Long Interview*, Grant McCracken advises that, when designing a research path, quantitative research will nearly always be preferred by the scientific community over qualitative methods. They are perceived to be more verifiable and more statistically useful than qualitative studies, and are generally less time-consuming and less costly to implement. 'As a rule of thumb . . . the research manager should treat quantitative methods as the default method. The question to ask is this: "Can I answer my questions using quantitative methods, or does the nature of my project require me to use qualitative methods?" The presumption of utility should always lie with the quantitative methods' (*ibid*,, 59). On the other hand, the researcher should turn to qualitative methods when 'he or she suspects that the issue at hand turns in some important way on the ways in which individuals conceive of, or construe their world' (*ibid.*, 59). Clearly, this applies to investigations such as the one reported in this book concerning the 18–24-year-old male drivers.

On the surface, quantitative methods have the greater weight of 'apparent validity' in their favour. Filtering human social responses through statistical analysis is more reassuring from a traditional scientific perspective than a verbal analysis of verbal material gathered by a single researcher. But issues of validity in both areas of activity are more complex than can be resolved by simple 'either/or' decisions. Before describing problems in individual data-gathering methods, the next section will discuss general issues of validity in social research. All methods involve unique problems of validity that must be balanced in setting research objectives.

2.6.2 Validity

Concerns about validity in social research reflect the ultimate objective of arriving at 'truth' in one's findings. Truth in the social sciences is not the 'absolute truth' of philosophy or metaphysics, but involves resolution of three traditional values from scientific inquiry that are mutually exclusive in practice: precision, realism (also called naturalism) and generalizability (Brinberg and McGrath, 1985, 43).

Precision in social research is defined as the accuracy of the measurements taken and the degree to which a researcher is able to control situational variables in the research environment that affect behaviours and responses. Realism refers to social context and the extent to which the environment in which data is collected reflects normal circumstances under which a particular behaviour occurs. Generalizability refers to the degree to which data collected from individual participants in a sample are applicable to members of the larger population under study.

In most measurement situations, these three values will cancel each other to some extent. Measurements are designed to maximize one, or at most two, of these research goals, but will always maximize one at the expense of another. Laboratory methods, such as those used in experimental psychology, will maximize precision by holding environmental influences constant during data collection, but will do so at a considerable loss of realism in observing participants in their natural environment. Ethnographic observation in the field can yield a high degree of realism, yet lacks both precision and generalizability to a larger population because of the limited ability of a single researcher to control situations or observe more than a few individuals in depth. Survey research could be highly generalizable to a population by virtue of a large sample size, yet fail to control the context in which data is gathered and variations in the way that the material is interpreted by participants.

These limitations are inherent in the methods themselves and cannot be overcome by improving any one procedure. The limitations revolve around the issue of scope. Methods that pursue broader scope or applicability will be accompanied by higher levels of noise, or factors that pollute the clarity of findings. The scope of methods that seek greater depth or precision will be more limited. Researchers in both quantitative and qualitative areas increasingly advocate a pluralistic approach to data gathering to gain perspective from the competing attributes and limitations of individual methods.

Kirk and Miller, in *Reliability and Validity in Qualitative Research,* observe that models for ensuring validity drawn from the pure sciences have, in recent years, come under scrutiny in the social sciences. Most experimental and quantitative research is based on the hypothetico-deductive model whereby a well-designed experiment strengthens or weakens assumptions held by the researcher. But in human research, social scientists have grown increasingly aware that 'hypothesis testing is appropriate to only a small proportion of the questions they ask' (Kirk and Miller, 1956, 17).

Questions formed in the mind of a researcher about a study population will always be formed from the social perspective of the researcher and might embody assumptions that do not reflect the reality of the study population. In these circumstances, it becomes irrelevant how precise a measurement is if the question asked contains incorrect assumptions. Qualitative methodologies have evolved in social research specifically to address this kind of error. Qualitative research attempts to frame questions that make sense in relation to the world-view of a study population by first identifying the components of that world-view.

Kirk and Miller suggest that problems of validity in qualitative research are primarily problems of objectivity. As they define it, objectivity does not imply an objectification of subjects or isolation of their behaviours from the human context, but rather a willingness on the part of the researcher to accept intellectual risk (*ibid.*, 10). The qualitative researcher adopts a posture whereby hypotheses that are generated can be proved wrong. The traditional confirmatory role of hypothesis testing in research ignores that many past significant scientific discoveries have occurred through accident or errors committed during methodological experiments. The qualitative researcher, therefore, intentionally enters into a study environment whereby unexpected events are sought and pre-existing assumptions are as likely to be weakened or strengthened.

Objectivity in qualitative methods involves resolution of two traditional research questions: reliability, and the original question of validity (*ibid.*, 19).

Reliability in research is the extent to which the same results are achieved every time a procedure is implemented. Quantitative social science has been heavily based on reliability in measurements, notably through refinements in sampling techniques and retesting strategies; but the certainty with which particular answers are obtained does not confirm their correctness. (The analogy of a thermometer that consistently measures the boiling point of water at 82°C is drawn.) Reliability assures that results are not accidental, but does not prove them valid. Validity, on the other hand, is the extent to which a procedure yields the correct answer. Most non-qualitative methodologies do not yet have systematic checks for validity.

Validity, as it is assessed in research, is of three kinds: apparent validity, instrumental validity and theoretical validity (perfect validity is theoretically unattainable). Apparent validity is a seeming truth upheld by 'the appearance of things', or something that is generally held to be true by its prevalence. Instrumental validity is a truth arrived at by alternative means in which the measure taken by an instrument B confirms measures taken by an instrument A. Theoretical validity is obtained by amassing sufficient evidence to suggest that the conceptual paradigm used to explain phenomena corresponds to observations.

In *Validity and Reliability in the Research Process*, Brinberg and McGrath examine validity issues in relation to the entire research process. They consider that the social sciences have attached too much importance to individual findings produced by individual methods, regardless of the precision with which the methods and appropriate verifications were applied. They suggest that validity in a study can only be assessed through examination of the original assumptions under which it was organized, and by careful analysis of the 'fit' of interrelated choices made by the researcher regarding theory, methods and the social phenomena under study to the stated purposes of the research (Brinberg and McGrath, 1985, 69).

Brinberg and McGrath suggest further that too little importance is attached to research failures as sources of useful information. They conceive the research process as a cumulative effort to reduce uncertainty with regard to a set of assumptions about specific phenomena, and consider that confirmation and failure to confirm hypotheses are equally valuable contributions to the field. They observe that these failures are both underreported and underused in analysis of qualitative data and consider that significant gains in information are subsequently lost (*ibid.*, 127). This cumulative definition of research is what Kirk and Miller refer to when they propose that scientific 'truth' emerges not as theoretical wholes, but as 'incremental, partial improvements in understanding' (Kirk and Miller, 1956, 12).

Generally, it is now accepted that validity in qualitative research requires multiple methods of data collection, and comparison, or 'triangulation', of findings across the methods. Validity also assumes a level of consciousness on the part of the researcher of his or her own biases in academic theory, methods and the social phenomena under study. Central to the ability of researchers to exchange and examine findings is the explicit statement in a published study of the theoretical and methodological assumptions under which a study is organized and the professional biases that the researcher brings to it.

Discussion of the data collection methods that follows reveals the different strengths and weaknesses of each regarding issues of precision, realism and generalizability to larger populations. These factors have specific implications for the nature and degree of validity that can be obtained from each method.

2.6.3 Participant observation

Participant observation, or ethnographic observation, is the original qualitative data-gathering methodology in the social sciences. The procedures that now characterize ethnographic research were developed early in this century by anthropologist Franz Boas, during his observations of eskimo villages, and refined by students and colleagues, including Margaret Mead, Ruth Benedict and Bronislaw Malinowski.

The distinguishing feature of ethnographic research is the immersion of a researcher in the culture under study for no less than one year. Although the method originated with studies of other world cultures, similar principles are now being used in studies of organizations and institutions in Western societies. The intent of the participant observation residency is that the researcher becomes deeply enough involved with the culture to discover how members of the culture think about, feel and understand events in their world while staying far enough removed to record the observations objectively (McCracken, 1988, 8).

The researcher is the main instrument for data collection in ethnographic research. The skills required include an anthropological theory of culture, empathy, perception of social and cultural nuances, and the ability to discriminate between large patterns of behaviour and idiosyncratic ones, as well as an ability to take notes, organize them and condense them into coherent form for analysis.

There are three main theoretical branches in ethnographic research: holistic, semiotic and behaviouristic. The holistic position holds that elements of a culture cannot be isolated from the entire cultural context, and it is concerned either with description and interpretation of individual cultures or with the idea of defining universal properties of culture. The semiotic position holds that the objective of ethnographic research is to determine both the meanings of cultural phenomena and how members of a culture understand the symbolic and organizational features of their world. Behaviourism in ethnographic research is concerned with the study of cultural particulars and often involves cross-cultural study of a single phenomenon (Sanday, 1983, 34).

Because of the commitment required from the researcher, participant observation research is the most expensive and time-consuming of data collection methodologies. Historically, ethnographic data analysis has been subject to criticism of bias because it frequently derives from the experience of one observer and is viewed as especially subjective. Ethnographic research is now increasingly performed by teams of researchers rather than individuals, and might include other forms of data gathering and testing during the study to verify findings as they are gathered.

2.6.4 Focus groups

Focus group interviewing grew out of techniques developed by clinical psychiatrists for use in group therapy. The principal theory behind interviewing in groups is that people are willing to divulge more sensitive personal information in the security of a group of strangers than they are in a one-on-one conversation with an interviewer (Bellenger *et al.*, 1976, 13). Also, the dynamic of group discussion sparks connections in the minds of participants that might not otherwise come out (much like the process of brainstorming for idea generation). It has also been observed that, because attitudes and opinions are naturally formed during social interaction, the group discussion provides an opportunity to observe actual processes of attitude formation. The group discussion also provides clues to participants' natural vocabularies on the subject because responses are usually less carefully formulated than in an interview. Where participant decision-making is important to the research question, the focus group can reveal the process whereby people arrive at beliefs. People often cannot explain how or why they hold certain beliefs, but interaction with others in the group sometimes provides clues as to how they reason through abstract ideas and propositions. The primary benefit of focus group research is therefore perceived to be the dynamic of the group setting itself, especially in studies where respondent attitudes are important to the researcher (Goldman, 1962, 44).

Focus group research is a controversial area within social sciences data collection. It is the primary method of qualitative research used in marketing, but is little used in the traditional branches of the social sciences. Both the benefits and disadvantages of the research method stem from the group environment in which data is drawn (Morgan, 1988, 12). Criticism of focus group research centres on the influence that the presence of others might have on responses offered in the group setting. Stronger speakers will influence comments by other group members to greater or lesser extents, and the opinions of one or two people, if not controlled by the moderator, can dominate a transcript. Focus group samples are so small (usually around 30–36 participants interviewed in three groups) that the results are not quantitatively significant, although the testimony is often so vivid that researchers and clients are tempted to treat them as generalizable to the larger population.

It is important for a researcher to remain aware that everything a respondent says in a group setting is 'self-reported' data (this is also true in private interviews). In other words, people's responses will reflect what they want others to think that they believe (rather than what they truly believe), and in many cases will reflect what they might want to believe, but do not.

Validity in focus group research hinges on a clear understanding of what can and cannot be gleaned from the sessions, and on the 'assumption that a measure really measures what it purports to measure' (Goldman, 1962, 46). There are well-defined standards for capturing 'cleaner' data from a session, but validity in a study will depend more on the analysis quality, for which there are fewer and less clear guidelines.

The group and the setting

The environment in which a focus group session is run is considered highly important to the quality of the results. The room in which the session is held should be as comfortable as possible. Living-room-like settings are ideal; imposing formal or laboratory-like settings are the least desirable. There should be no obvious two-way mirrors or video cameras in the room, and respondents should be informed of them if there are. Observation and recording should be as unobtrusive as possible. Focus group sessions are always audiotape-recorded because respondents generally feel more certain that their anonymity is assured this way than with video.

Participant selection is also considered central to assuring usable data. Group members should be recruited from as homogeneous a population as possible, for the obvious reason that people will open up more to a group of strangers that they perceive to be 'like them' than to people from different backgrounds. Similarity in sex, age and income tend to be the most crucial factors in ensuring a comfortable exchange within a group. Race is a factor only in locations where strong racial tensions exist. Age and ethnicity can be factors to consider if the group is supposed to represent a highly specific segment of the population (Axelrod, 1975, 79).

The skill, preparation and temperament of the moderator also influence the quality and reliability of data. The moderator must be able to convey objectivity, and to ensure that any biases, whether personal or those of the research team, are not expressed in the session. The moderator affects an attitude somewhere between empathy and detachment; he or she should seem pleasant and interested in all comments, but remain absolutely neutral as to the discussion content.

It is helpful if the sociographic profile of the moderator is similar to that of the group; however, given the specialization of moderator training and the range of potential study populations, this is often difficult to arrange. If the topic under discussion is at all sensitive, it is imperative that the moderator be of the same sex as the group, because men and women talk differently among themselves than they do in mixed company, and the objective is to capture a conversation between peers. If the moderator is of the opposite sex (and especially if that person is attractive), group members will perform for the moderator rather than talk among themselves. It is also important that the vocabulary used by the moderator be comfortable for the group.

Running the sessions

The moderator should be well prepared going into the session to ensure that important topics are covered in depth and to manage the dynamics of the group. It is the responsibility of the researcher (or research team) to provide the moderator with background information in the topic area if he or she is not already familiar with it. The researcher writes a thorough set of questions and research objectives to discuss with the moderator days in advance of the sessions. Research objectives should be reduced to one or two key questions that will open up various topic areas to be discussed by the group. The moderator should have this 'script' of questions more or less memorized so that the shuffling of notes or cue cards is minimal. A group will not always wander naturally onto subjects that the researcher wants addressed, so the moderator should be prepared to steer conversations towards these points without implying that particular responses are being sought. One of the benefits of group interviewing is that a sensitive issue can be raised by a group member and discussed at length, whereas it would be highly threatening if asked directly by the moderator.

Different levels of moderator involvement are appropriate for different research applications, but a low level of moderator involvement is generally preferred for social sciences research. Respondents should not feel that they are answering the moderator so much as talking among themselves. This atmosphere provides the greatest naturalism and richest interaction (Morgan, 1988, 32).

Questions can be introduced by the moderator that minimize self-presentation in respondents' remarks. These 'projective questions' focus away from personal preferences (which will always be face-saving) and onto underlying assumptions through which attitudes and opinions are formed. Rather than asking 'What do you think of this car?', the question 'What kind of person would want to buy this car?' is more likely to reveal cultural categories and logic used by the respondent. 'Deprivation questions' such as 'What would you miss most if you were no longer able to...?' can put respondents in a reflective mood and enable them to think abstractly about otherwise mundane topics (Goldman, 1962, 46).

Opinions offered out of context are dangerous, because a researcher never knows where and how the opinion was formed, or whether the opinion is truly held by the respondent. A technique for both verifying an opinion and stimulating discussion is occasional prompting with questions such as 'What experiences have you had that lead you to feel that way?' The more the respondents talk about actual experiences, the richer the final transcript will be in categories and logical constructions (Axelrod, 1975, 56).

It is important, however, for a moderator not to 'overprompt', as people quickly learn that speaking up might mean being asked to divulge more than they are willing to share.

Analysis and reporting

Methods of analysis in focus group research are not as well documented as are the standards for running the groups. Some researchers prefer to report only what was said and leave interpretation of the material up to the report reader. Others suggest that it is important to analyze 'what happened' in the group as well as what was said. This approach includes exploring the character or tone of exchanges between participants, and notation of topics that were avoided or dropped quickly and those that were pursued enthusiastically. Participant choices during the session reveal the prominence of certain issues over others (*ibid.*, 56).

Like other forms of qualitative data, information gleaned from focus groups cannot be considered statistically significant. Focus groups are most often used in combination with other data-gathering methods, especially where insight is needed into a new topic area from the perspective of the study population. Focus groups are used as exploratory tools for generating questions, categories and appropriate vocabulary to be tested in quantitative questionnaires. They also help to discard or confirm assumptions derived from other sources and frame the issues of a quantitative research questionnaire.

2.6.5 The interview

There are several different kinds of interviews in social sciences research that serve quite different purposes as information-gathering tools. Three important versions are the ethnographic interview, the depth interview and the long interview.

The ethnographic interview is an intensive method used in field anthropology which depends on immersion of the researcher in the culture under study. The ethnographic interview takes place over a number of sittings with the objective of achieving as much environmental naturalism as possible by allowing time for conversations with the researcher to seem like a relatively normal part of the setting. Ethnographic interview data is descriptive of the cultural values and behaviour of individuals within specific social structures and is often used to support participant observation techniques.

In sharp contrast, the depth interview is more akin to techniques used by clinical psychologists that yield a comprehensive psychological profile of an individual. The process is primarily concerned with identification of the personality factors and affective states that cause specific behaviours in one person, and is not intended to be generalized to any class of subjects without mediation by other measurement tools.

The long interview is a more economical form of ethnographic interview designed for use by researchers working within their own cultures. Interviews with individual participants can serve as an important balance to focus group interviews in that they eliminate the influence of other people (other than the interviewer) on a respondent's testimony.

Members of North American and European cultures lead private lives, and most people will not, or cannot, agree to being observed or interviewed for extended periods. It is extremely difficult for a researcher to enter the workplace for interview purposes, and it is always a delicate procedure to gain access to a private home. Time constraints further prevent people from being able to commit to lengthy participation in a study, even when the privacy barrier has been bridged.

The long interview, as defined by Grant McCracken in *The Long Interview*, provides a structure for revealing the mental world of individuals within a particular culture. The technique is intended to expose how a person defines events and objects pertaining to the topic, understands his or her experiences with it, and reveals the logic by which decisions and choices are made.

The long interview allows enough time to be spent with the respondent (six to eight hours broken into two sittings) for primary constructs to be exposed without unduly taxing the respondent. Planning methods are employed that enable the researcher to recover key data economically from the large amount of raw material collected.

Through an extensive literature review, the researcher becomes familiar with existing assumptions about causes of behaviour in the subject area and makes lists of categories for investigation. From potential categories, the researcher identifies areas that might be expected to yield greatest insight into how the respondent understands the issue.

Based on the above, and on a systematic examination of his or her own experiences with the topic, the researcher develops definitions and assumptions about the issue and formulates a set of unanswered questions. The researcher creates a preliminary set of expectations about what the respondent's views might be with the understanding that they could be strongly contradicted during the interview phase.

The next step involves writing a questionnaire to use as a template for the interview. The long-interview questionnaire differs considerably from those used in door-to-door and telephone surveys. Rather than a list of carefully worded questions, to which a respondent supplies answers for comparison with those of other respondents, the long-interview questionnaire is a flexible series of open-ended questions and supplemental prompts. It functions like a 'guidebook' for a journey into the respondent's thought processes about the topic. The long interview is not concerned with sample size and seeks to obtain depth of information from only a few respondents (usually around eight). In contrast to focus group research, which assembles as homogeneous a group as possible, the long interview study seeks respondents with demonstrably different profiles from within the identified population.

The questionnaire begins with short biographical questions to give the respondent an opportunity to become comfortable with the interviewer and the process and to clarify precisely who is being interviewed at the head of the transcript. The interviewer should establish a relationship with the respondent that strikes a balance between personability and professionalism. If too much rapport is created, the interview will lose its focus on the respondent's testimony and the respondent will attempt to engage the interviewer in a two-way conversation, try to please the interviewer, or withhold potentially embarrassing data, as would be typical in a newly formed friendship. If too much professional distance is established the respondent will be reluctant to reveal experiences or personal thoughts and might simply try to get through the interview in a face-saving way.

The respondent will naturally try to analyze the interviewer, even as he or she is being interviewed, and will assess the level of personal risk involved. It should be clear at the outset that the interviewer is a neutral, pleasant person who is willing to listen with interest to everything that the respondent has to say without judgement. If a comfortable tone is established, respondents often find the process stimulating and will use the opportunity to express themselves fully without fear of ridicule or social consequence.

The way in which questions are posed is central to the quality of the interview data. The purpose of the interview is to allow the respondent to tell the story in his or her own terms, and the interviewer must take care to record logic and categories as they are supplied without imposing interpretations on the experiences being reported. The questioner should not engage in 'active listening' (a technique used in counselling and analysis), whereby the listener restates in other words what the speaker is saying. Any commentary by the interviewer can influence the direction and character of the respondent's testimony.

The interviewer should play 'dumb', asking only for more depth or clarification, such as 'Can you tell me more about that?' or 'What do you mean by the word "_______"?'

The interview consists of a few large questions which the respondent is encouraged to talk about at length. Once a principal question is asked, the interviewer's role becomes one of administering prompts that keep the respondent talking in ever greater depth, but in the general direction of the questionnaire. Prompts consist of non-verbal signals of acknowledgement, such as raised eyebrows, nods or sounds that register interest and comprehension and keep the respondent talking. The researcher must stay alert to the narrative paths that a respondent chooses when making a point and judge whether the discussion is unproductively off the topic or a unique way in which the respondent constructs logic.

In formulating questions, the researcher accepts that respondents sometimes have difficulty explaining actions or behaviours that are so habitual that they have never had cause to be introspective about them. Questions about unusual events in the area of the topic often have a revelatory effect. Skilful questions and prompting can cause the respondent to view an experience in an entirely new way while reporting it, thereby enabling the interviewer to record the process by which the respondent assigns meaning to the event.

In the prompting process, the interviewer can lead the respondent towards deeper definitions of categories with 'contrast questions', for example, 'What is the difference between X and Y?' It is important for the interviewer to know in advance what must come out of the interview and to prompt for it without influencing the response.

2.6.6 The survey questionnaire

The survey questionnaire is a quantitative data-gathering instrument, and it is frequently used in combination with qualitative methodologies to verify findings of many kinds. The usual reason for including a survey in a qualitative study is to measure the degree to which assumptions formed during contact with a small sample can be generalized to a larger population.

There are two kinds of question in survey research: questions that deal with facts or behaviours and those that pertain to attitudes or psychological states. In principle, facts or behaviours are verifiable because they can be observed; however, the notion of tracking a large sample of respondents to observe reported behaviours would be absurd in practice, especially if the behaviour were a private one. Questions about attitudes or psychological states, on the other hand, are inherently unverifiable because they cannot be observed. The problem for the survey researcher then is to be able to establish with some confidence that answers supplied by respondents are 'true' (Sudman and Bradbum, 1982, 19).

Truth in survey data is vulnerable to a range of sources of human error that can be organized under three questions:

1. Has the respondent answered truthfully? (The subject ignores the answer, or has not thought about the issue, but responds anyway to avoid appearing uninformed or unaware.)
2. Does the response truly represent how the respondent thinks or feels? (The subject thinks in one way but responds in another, perceiving some form of risk in answering truthfully.)
3. Has the respondent interpreted the question correctly? (The subject misunderstands the question and answers truthfully or not to the question as he or she understood it.)

Good analyses of this topic can be found in Converse and Presser (1986), Platek *et al.* (1985) and Sudman and Bradbum (1982).

Asking questions

A number of guidelines have evolved during the last 50 years for minimizing error in standardized surveys. Converse and Presser, in *Survey Questions: Handcrafting the Standardized Questionnaire* (1986), stress the values of clarity, simplicity and intelligibility in writing a good survey to avoid overburdening the respondent. They observe that questionnaire writers are usually highly educated and deeply involved in the jargon and academic minutiae of a topic area. It is extremely important for a researcher to step back from the process and view the topic and the survey from the respondent's point of view. First drafts are usually too long and too wordy, and include tasks that are too complex for a respondent to want to negotiate on volunteer time (*ibid.*, 9).

The best way to eliminate the most common source of error, a misunderstood question, is to use simple language that is used by most people. The vocabulary should be as basic as possible, that is, 'don't say "principal" when you can use "main"' (*ibid.*, 9). The researcher should be sensitive to the way that questions are asked in normal conversation, without being too familiar or using slang. People are quick to recognize artificially casual tones, especially in official-looking documents, but wherever a correct grammatical construction sounds too stiff, a choice should be made in favour of less correct but more natural phrasing (*ibid.*, 9).

A means for detecting whether people have guessed at answers, misunderstood them or made choices that do not reflect their true thoughts or values is to build several differently worded questions on the same topic into the questionnaire for cross-checking. Conflicting answers on questions that the researcher believes ask the same thing are a clue that something is amiss.

Another tactic is the use of 'don't know' or 'I feel neutral' categories that give the respondent the option to state that he or she either does not know or has not formed an opinion on the subject as an alternative to guessing or making up answers. Researchers disagree on whether the neutral category is necessary or not. Some consider that forcing people in one direction or the other on opinion surveys is more valuable than allowing them to hug the middle without committing. Others feel that it is more significant to know that the respondent holds no strong feelings about a topic than whether they lean slightly more in one direction than another.

A common first-draft error in question design is the double-barrelled question, or a question that has two parts to it that the researcher has lumped together. Double-barrelled questions usually harbour the words 'and' or 'or' in ways that the researcher might not be alert to; for example, 'Which source do you use most for news and information?' (News and information are two quite different things, although they are often used interchangeably in media jargon.) The use of double negatives is also to be avoided; in other words, the word 'not' should not be used (*ibid.*, 13).

Another problem in survey results is the tendency of people to agree with statements just to be agreeable. This problem increases as income and education levels drop. To combat this tendency it is important to make clear both sides of agree/disagree questions: 'Do you agree with the government's economic policy?' will draw more agreement than if 'or not' is added to the question. Generally, agree/disagree questions should have some kind of intensity scale built into the question or immediately following. People prefer to have more than two or three choices when expressing opinions, and determining whether opinions are strongly or weakly held is usually central to obtaining a meaningful data analysis.

Closed-ended vs. open-ended questions

A major decision to be made in survey design is when and where to use open-ended rather than closed-ended questions. Open-ended questions are those for which no predetermined choices are provided; the respondent is free to offer whatever comes to mind in response to the question. Closed-ended questions are those for which a specific number of categories or choices are prepared from which the respondent is asked to choose. Open-ended questions are associated with qualitative inquiry; closed-ended questions are more suitable for quantitative measurement (Platek *et al.*, 1985, 58).

There are advantages and disadvantages in using either open-ended or closed-ended questions that are largely determined by the purpose of the question and the uses to which the gathered data will be put.

The advantage of open-ended questions is that the format does not place boundaries on the possible responses that a participant can give. The respondent cannot guess at an open-ended question, and the response usually reflects what the respondent considers most salient in the topic area.

The disadvantages of open-ended questions are that they place more burden on the respondent than closed-ended questions and are much more difficult to codify and analyze statistically. Open-ended questions can offer an opportunity for self-expression, but the time and involvement factor could just as easily turn a respondent off. This is especially true in mail surveys where writing tasks might be perceived to be too difficult, or too time-consuming, and the respondent will simply refuse to finish the survey. Reticent people will be less inclined to answer at length than outgoing people, even though they could be just as knowledgeable. Colourful responses will attract more attention during analysis than more ordinary accounts – even though they could be no more significant – and can bias the conclusions. Coding open responses for analysis is often tedious and requires judgement and familiarity with the research plan on the part of the coder. Use of a codifying process might also suggest that open responses can be analyzed statistically like closed-ended responses, but this might not be true.

The main advantage to closed-ended questions is that they are easy to codify and tabulate, and assuming that validity problems have been resolved in the writing process, they can be analyzed statistically. All respondents in a closed-ended survey answer the same questions and choose from the same set of responses, eliminating bias that could come from vividness of testimony. Closed questions will often offer a range of responses that people find easy to choose from, but for which they would have difficulty supplying answers in an open format.

The chief disadvantage to closed-ended questions is that the format assumes that the question writer will be able to provide categories for all appropriate answers. If respondents cannot find categories to fit their true responses, they will leave the question blank, write something in the margin, or choose another answer that really does not match. All these choices produce invalid responses.

Closed-ended questions also lead respondents to frames of reference that might not reflect their usual orientation to the topic. It is essential, when designing closed-ended questions, that contact be made with the study population to determine the kinds of categories that they use to discuss and evaluate phenomena in the topic area. Qualitative inquiry, such as focus group interviewing, can provide valuable insights for writing closed-ended questions.

A fairly standardized process exists for planning and designing questionnaires that helps researchers to identify and deal with validity issues in advance of final printing and administration of the questionnaire. The first step is to define the research problem and to identify key information gaps remaining from the literature search. Planning begins with the following questions: 'What is the problem I am trying to solve?' 'What must I know to solve this problem?'; 'How will I use the information once I get it?'; 'How accurate does the information have to be?' (Sudman and Bradbum, 1982, 22).

The remainder of the design process can be outlined as follows:

1. The researcher states expected uses of the data.
2. The researcher states the need for the information in relation to specific research objectives.
3. The researcher writes a schedule and negotiates resources.
4. The researcher determines the composition of the research team, ideally including individuals with a background in mixed disciplines, especially in sociology, statistics, the behavioural sciences, and those with practical experience in the topic area.
5. Research team agrees on study objectives, acceptable levels of data quality, sample size, etc.
6. Review of literature on existing questionnaire designs used in similar studies.
7. Draft initial questions, share and discuss with colleagues, meet and pretest with groups similar to the study population for language, comprehension and logic.
8. Write and format survey. Questionnaire designer plans logical groupings.
9. Share with colleagues, rewrite, edit and refine draft questionnaire.
10. Pretest questionnaire with sample study population;
11. Rewrite, edit and refine pretest questionnaire.
12. Assign codes to specific questions.
13. Produce and administer final questionnaire.
14. Decode open questions and prepare data for analysis.
15. Statistical analysis.
16. Analysis by researcher for drawing conclusions.
17. Publication of research findings.
18. Observation of problems noted along with recommendations for future questionnaires (Platek, *et al.*, 1985, 19).

Questionnaire design: the introduction

A survey should always be preceded by some kind of introduction that clearly states the purpose of the research and tells the respondent how results from the survey will directly benefit him or her, that is, through increased social welfare. Ethical considerations require that the respondent be given enough information about the purpose of the study and uses to which data will be put to make an informed decision whether or not to participate. Assurances of confidentiality should be included, but only to the extent that confidentiality can be guaranteed. The amount of information given to the respondent about the purpose of the study is generally proportional to the 'risk' incurred in supplying the requested information (Sudman and Bradbum, 1982, 9). The introduction can be a key factor in influencing the respondent's willingness to take the time (or assume the risk) to participate.

Demographic data

A certain amount of demographic information about the respondent is usually required to analyze collected data meaningfully. If the survey is administered by an interviewer, a set of non-threatening biographical questions can be used to open the survey. This permits the respondent to become comfortable with the process and establishes some descriptive personal data up front. Demographic questions that are in any way threatening, however, should come at the end of the survey. Questions about income, marital status, age and religion are perceived by some people to be highly personal, and if they appear too early in the sequence of questions the respondents might refuse to finish the questionnaire. Respondents might choose not to answer sensitive questions that come at the end, or will lie about them, but by then all other pertinent questions would have been answered. Ending a survey with a quick set of demographic questions will usually seem easy and come as a relief to respondents who have just finished more time-consuming or difficult questions (Converse and Presser, 1986, 63).

Flow of the questionnaire

There are certain established guidelines for planning and designing the flow of questions in a survey that make the task easier and more pleasant for the respondent. The questionnaire should begin with questions that are easy and fun. If initial questions are difficult or boring, the respondent will lose interest and quit. Questions should move from the general to the specific, like the natural flow of questions in a conversation, and be arranged in logical groupings, preferably introducing each group with a brief unifying statement. If respondents perceive that questions jump around or observe illogical patterns, they could become annoyed, suspicious or fatigued trying to anticipate what comes next. Instructions for each section must be absolutely clear.

The layout of printed, self-administered questionnaires should be clean, professional-looking and as easy to fill in as possible. Printed surveys are best folded and saddle-stitched, rather than stapled at a corner, because the format is simpler for the respondent to manage and seems less imposing. Pages in this format can be easily flipped and scanned before the respondent begins the questionnaire. Multiple-choice options to a question should always be aligned vertically, because horizontal arrangements cause confusion as to whether boxes precede or follow an answer. The printed survey should contain as few pages as possible, ideally no more than four for a mail survey. Crowding questions together to save space is not a good alternative, because dense pages will make the survey look more difficult to complete than a spacious one with more pages (Platek *et al.*, 1985, 93). Finally, the questionnaire should always end with a statement of thanks.

2.6.7 Methods of analysis of qualitative data

Less has been written about methods of analyzing qualitative data than about their collection. Typically, textbooks on qualitative methodologies devote fewer than 5–10 per cent of their pages to discussions of analysis, and many completed studies are published without a detailed reporting of the methods used during data analysis (Miles and Huberman, 1984, 16).

Qualitative research is much younger and more controversial than research in the physical sciences, but its origins in anthropology have left analysis largely up to the individual researcher. The tradition has been subjective to the point that some in the field have declared qualitative analysis an 'art' that cannot be learned (*ibid.,* 16). In relation to problems in qualitative analysis, Mathew B. Miles states:

> The most serious and central difficulty in the use of qualitative data is that methods of analysis are not well formulated. For quantitative data, there are clear conventions the researcher can use. But the analyst faced with a bank of qualitative data has very few guidelines for protection against self-delusion, let alone the presentation of unreliable or invalid conclusions to scientific or policy-making audiences. How can we be sure that an 'earthy', 'undeniable', 'serendipitous' finding is not in fact *wrong*? (Miles, 1979, 117)

Nevertheless, researchers agree on certain points that are central to shared notions of quality in analysis. These include thorough, methodological systems for data transcription; methodical data review and sorting; reduction of data; separation and analysis of key ideas; display of phenomenal relationships; and elegance of summary.

Researchers agree that a point of great importance to guaranteeing accuracy and validity in analysis is the creation of an accurate transcript. The transcription of interview tapes should be done by a professional transcriber. The researcher should not perform this task since overfamiliarity with the material and conscious or unconscious self-editing can lead to error.

Grant McCracken (1988) describes in detail a five-step data-analysis process for organizing the raw material of data collection. The process is akin to other 'intuitive' systems of analysis and is applicable to most qualitative research problems. McCracken stresses creation of a 'paper trail' during analysis, that is, a visible record of the sorting and evaluation decisions one makes that permits another scholar to retrace the thought processes that led to conclusions. This physical record is crucial to ensuring that the study can be cross-checked by others for reliability and validity.

In the first phase of analysis, the transcript is read objectively with no attempt to link ideas or generate interpretations, but with a conscious eye for significant 'utterances'. These remarks are visually or physically separated from the body of the text. The utterances are studied in relation to the immediate context in which they appear in the transcript, without reference to other concepts or statements. The researcher lists all possible meanings that could be implied by each individual statement, that is, 'What else could the respondent be saying here?' Utterances become surrounded by branching notations of possible meanings associated with the statement. During this phase, the researcher exercises discipline in not jumping to conclusions about the larger significance of the utterances.

In this first phase of analysis the researcher should take metaphors at face value. Common metaphors that the respondent does not necessarily think about when speaking often have their origins in physical experiences that, in fact, shape the way that the respondent views an experience or attaches later meaning to it. 'When the respondent speaks of being "blasted," it is worth taking this figure of speech at its face. Is the person violently and suddenly destroyed by heavy drinking? Plainly not. Well, is the social person in some sense so destroyed? Perhaps so' (*ibid.*, 44).

In the second stage of analysis, the researcher begins to match utterances in the transcript with concepts from the literature review to form a set of observations. Using existing ideas about behaviour in the topic area, the researcher compares utterances from the transcript against existing assumptions to see where the data conforms to established thinking and where it does not. At this stage, the researcher also compares isolated utterances against other utterances in the transcript for consistency and contradiction in the respondent's testimony.

In the third stage, the researcher explores relationships between utterances with an emphasis on observations made in the second stage. Actual utterances become less important than observations and the possible meanings drawn from them. At this point, a field of patterns in the respondent's constructed world should become apparent.

In the fourth stage, the researcher begins to make judgements. Major themes are identified, and relations between themes noted. One or two primary themes are established under which all other points are organized. These first four steps are completed independently for each interview conducted in the study.

The fifth and final stage involves the formation of a thesis by comparing relations between themes identified across the interviews. The thesis becomes the set of conclusions for the study.

Miles and Huberman (1984) describe a large number of similar paths of analysis for specific types of applied problems in qualitative research. Their methods are designed for larger studies that include multiple-member teams working across numerous research sites with an emphasis on problems in ethnographic data. These methods are designed to ensure that all members of the research team follow roughly the same procedures for sorting, coding and signifying data, and that reporting of interim conclusions occurs throughout a project.

Miles and Huberman have made a significant contribution to the area of qualitative analysis by advocating data-display techniques as fundamental tools for discovering relationships within collected data. The traditional narrative format for reporting qualitative observations and conclusions is cumbersome and tends to bury phenomenal relationships in descriptive detail. By forcing analysis to assume simplified, visual expression, through matrices, diagrams and charts, and other means of display, researchers can quickly ascertain significant relationships in material gathered during fieldwork. The resulting compact diagrams are also easier to read quickly, comprehend and store, a factor that makes them more accessible to others working with similar problems.

2.6.8 Conclusion

A survey of data collection methodologies in marketing and the social sciences quickly reveals that questions of validity and objectivity are primary concerns for anyone seeking information directly from a study population.

Validity can be considered to be the degree to which research findings accurately reflect the perspectives of the group under study for a particular phenomenon. Objectivity refers to the extent to which a researcher acknowledges theoretical and cultural biases that he or she brings to the process of inquiry, and potential and inherent limitations in the methods used.

Establishing validity in a study involves recognition that every data collection method has specific benefits and limitations in the kind of information that it can yield. A primary goal of all research is to maximize certainty regarding the precision, naturalism and generalizability of findings, but it is clear that any method will maximize only one or two of these values at a time, and will always do so at the expense of the third. Given the inherent limitations of individual methods, the best way to ensure that the results of a study reflect what they are intended to reflect is to approach data collection from multiple angles, using methods that complement each other in strengths and weaknesses.

Data-gathering methods are of two primary kinds, defined by two different research questions. Quantitative methods measure the degree to which certain attitudes or behaviours are manifest in a study population. Qualitative methods reveal the psychological, social and cultural motivations that underlie them. Although quantitative methods have been used more extensively in social sciences research, qualitative methods are increasingly being used to frame and define questions for quantitative measurement.

In a situation of pure inquiry into an aspect of reality, a social scientist could choose to stop in front of uncertainty, in the hope that future developments in quantitative methods would allow shortly an extended measurement of reality. A designer, however, must act in the construction of communications, and has to consider any method that would assist in the search for better decision-making bases. Given the complexity of human response to communications, the designer has to recognize that restricting the information about the public to only those dimensions that can be quantitatively measured would leave too much out. Seemingly, the best option is a combination of methods, where the composite picture provided by them, despite their individual limitations, would offer a base on which an initial approach to the design of a communication strategy can be grounded.

Because the purpose of the study of 18–24-year-old male drivers referred to in this book is to identify the attitudes and motivations that most strongly influence the specific decisions that they make while driving, we determined from the beginning that a qualitative investigation was most appropriate to our problem (noting, however, that the choice of the segment of the public to be addressed was made on the basis of quantitative information). We identified focus group research as a starting point because of the advantages of group interviewing for flashing unanticipated connections and values, and for revealing the vocabulary and manner in which members of the group converse with each other about driving. We recognize that comments gathered in the group environment tend to be strongly characterized by self-presentation, but consider this information valuable in itself. How young men relate their driving experiences to others strongly reflects not only how they want to be perceived as drivers but also their driving culture, values and perceptions.

We consider the focus group results an important foundation for further identifying significant attitudes and behaviours among young adult male drivers that contribute to their overrepresentation in fatal and injury-causing motor vehicle collisions.

Isolating these attitudes and behaviours enabled us to begin to match specific driver errors with particular cultural beliefs to create messages that have implications for new kinds of behaviour role-modelling in driving.

We recognize that results from these initial focus groups are not generalizable to the general population 18–24-year-old male or to specific higher-risk groups within that population segment. We therefore designed a survey instrument for quantitative verification of attitudes brought to our attention in the focus group discussions. We anticipate that conducting long interviews with individual members of the study population would possibly reveal somewhat different and more personal perspectives than responses gathered in focus groups. Driving, however, is a social act, and although the individual interviews might provide different perspectives, this would not invalidate the issues raised in the group setting.

CHAPTER THREE

Targeting communications

Traffic safety project report

3.1 INTRODUCTION

Chapter 1 dealt with the generalities of a framework, concerned with communication and design principles, which is indispensable for the grounding of specific actions in design. Chapter 2 discussed design methods in general. This chapter reports on specifics, concerning the development of a strategy for a traffic safety campaign.

The objective is to present the details of a decision-making process, as a methodological model relevant for future action.

The work method reported, properly adapted, could be extrapolated and applied to other problems and cultures; however, this extrapolation must consider a great number of details that distinguish every problem and constrain any importing of methods and processes. There will be enough information in this report to make it useful for those interested in developing communications and programmes aimed at affecting existing public attitudes.

This chapter provides a model for a number of critical steps of the process:

1. Identification of a social problem to be addressed.
2. Identification of a specific segment of the public that significantly contributes to the existence of the problem.
3. Definition of the target population profile.
4. Definition of specific objectives, actions and a communicational strategy.
5. Definition of the verbal and visual arguments and structure of the communications to be developed.

3.2 CREATING A BACKGROUND

3.2.1 Identifying a problem

Many problems affect society today, and it would be difficult to agree on which is the worst, wrestling between the urgent and the important. From the designer's point of view, however, there are two questions to keep in mind when selecting a problem to address:

first, is the problem significant, and second, could visual communications make an important contribution to its reduction?

A description of problem significance is necessary to approach agencies that could finance first the research project, then the planning, and later the production, implementation and evaluation of the campaign. A detailed assessment of the magnitude of the problem, developed along with a proposal describing what could be done about it, has many purposes:

1. It assists the investigator in the construction of a definition of the problem in communicational terms.
2. It shows potential supporters the need to take action and proposes possible actions to be taken.
3. It demonstrates to agencies that the proponent knows the problem, has novel views about it, and offers a new approach to deal with it.
4. Last, and most important, it provides the designer with an entrance to the circle of institutions that play a role in the issue at stake, therefore providing access to new information, to specialists, to interaction at the level of paradigm-setting, and to future channels for action.

The traffic safety study on which this chapter is based was supported by the Alberta Solicitor General and the Alberta Motor Association, and was assisted by these agencies and by the contribution of marketing and educational psychology consultants at the University of Alberta.

The project

After defining the problem and the main target group for the supporting agencies, Dr Adam Finn (marketing) joined the team to contribute to the profile development of the selected target population (18–24-year-old male Alberta driver), particularly in connection with their driving attitudes and beliefs. This target population was selected because of its overrepresentation in traffic collision statistics. Finn suggested the organization of three focus groups, each one of about ten people. For these groups, he prepared a set of open-ended questions, moderated the sessions (where the author was present), recorded them and prepared a report with an analysis of the proceedings. A visual component was added to his proposed structure for the focus groups: the author showed the groups some alcohol and car advertisements and some safety posters to obtain reactions to visual materials that could assist when moving into the conception of actual visual communications.

After Finn's report was completed, two educational psychologists, Henry Janzen and John G. Paterson, prepared a target population profile based on existing literature, analyzed the focus group transcripts and the Finn Report, and made recommendations on the production of traffic safety communications appropriate for this target audience.

At the same time, Zoe Strickler – then a communication design graduate student under my supervision – wrote a paper on data collection validity, to frame the information gathered. Section 2.6 of this book is based on that paper.

After this was completed, I worked with the focus groups transcripts, the Finn Report, the Paterson/Janzen Report, the Strickler Report, relevant bibliography, my own notes and the research proposal to prepare the introduction, discussion and recommendations of *Traffic Safety in Alberta/Casualty Collision and the 18–24 Year Old Male Driver: Criteria for a Targeted Communication Campaign* (1992). The following material is based on that report.

3.2.2 Assessing the dimension of the problem

'Motor vehicle accidents are a major health and safety problem in our country today. In the last decade, 51 300 people died in road crashes in Canada; another 2 342 300 were injured – many of them [were left] permanently disabled. In fact road crashes are now the leading cause of death to people under the age of 45 and a major cause of injury' (Road Safety and Motor Vehicle Regulation Directorate, 1987, 2). In Canada, traffic injuries kill twice as many people under 35 as cancer and heart disease combined; however, whereas cancer and heart disease are areas with substantially funded research on prevention, little is dedicated to the study of strategies aimed at reducing the traffic injuries epidemic. In the USA, 'Injury research expenditures are estimated at $160 million for fiscal year 1987 compared with expenditures for cancer research by the National Cancer Institute of 1.4 billion. The National Heart, Lung and Blood Institute spent $930 million for cardiovascular research in fiscal year 1987' (Rice *et al.*, 1989, xxxvii). The difference between funds directed at cancer and heart research and those directed at traffic safety research is staggering, and it does not include the millions of dollars invested in research and development of cancer- and heart-related drugs by the pharmaceutical industry.

Figures 3.1 and 3.2 show clearly the importance of the problem: in intensity compared to other injuries and presence in many countries. They will also show the strong correlation that traffic crashes have with youth.

Wars always attract public attention, create heavy reactions and mobilize the press, but 'if every war since 1776 is taken together, no instrument of death – flintlock, repeating rifle, machine gun, tank, plane or bomb – has resulted in as many American fatalities as has the motor vehicle. From 1910 to 1985, there were more than 2.5 million traffic fatalities' (National Committee for Injury Prevention and Control, 1989, 118). In 1991, 43 500 people died in car crashes in the USA, and 1 600 000 suffered disabling injuries (an injury disabling the person beyond the day of the accident). Of these, 130 000 were left with permanent impairments (National Safety Council, 1992, 1).

It is impossible to quantify the human costs inflicted upon individuals and their families. Spinal cord, brain and other severe injuries mark people for life, reduce their potential contribution to society and often contribute to other future ailments. In most severe collisions, the health care and suffering do not end on the day of the crash.

In addition to the human suffering, the cost of traffic crashes is enormous. Their direct cost in the USA for 1991 was 96.1 billion, including medical expenses, wage loss, insurance administration, motor vehicle damage and uninsured work loss (*ibid.*, 2). Seen from the perspective of the economy, happiness or life expectancy, traffic collisions are a major burden for human life and public administration.

The present preliminary study proposes to reduce the problem through the development of an efficient and cost-effective communication campaign.

> A review of past campaigns makes evident that a long period of skepticism about the strategic use of mass media in health promotion was strongly influenced by findings from short-term studies that focused too narrowly on immediate behaviour change and were seriously flawed by design and execution. . . . There is growing recognition that [health promotion] campaigns more often succeed when they address an issue of ongoing public concern and incorporate both the 'consumer' orientation of commercial marketing and research-based principles of behavior change. (DeJong and Winsten, 1989, 6)

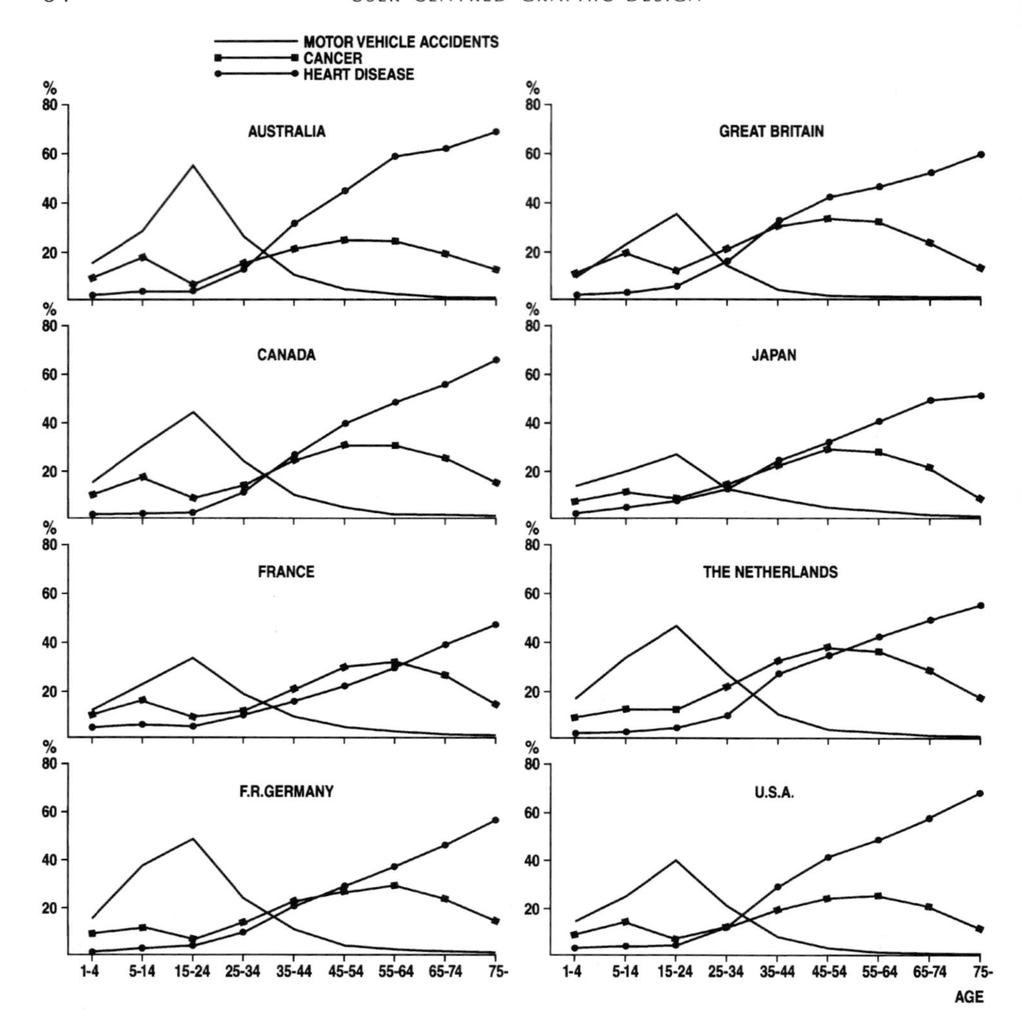

Figure 3.1 Percentages of heart disease, cancer and motor vehicle traffic accidents of all the death causes within each age group in eight countries in 1970. (Source: WHO, 1973, in Näätänen and Summala, 1976, 13.)

3.2.3 Defining a first-priority target group

Communication campaigns that target specific groups are more effective and efficient than generic campaigns (Näätänen and Summala, 1976, 102). In this case, to optimize the effect of resources used, it was decided to address the campaign to the sector of licensed drivers who generate the highest number of casualty collisions. (I use the term *casualty collisions* in this text as short for *casualty-causing traffic collisions* for the sake of brevity, and to avoid *traffic accident*, a term that transforms the errors of drivers – accounting for 97 per cent of collisions – into capricious events – *accidents* – that can happen to anyone.)

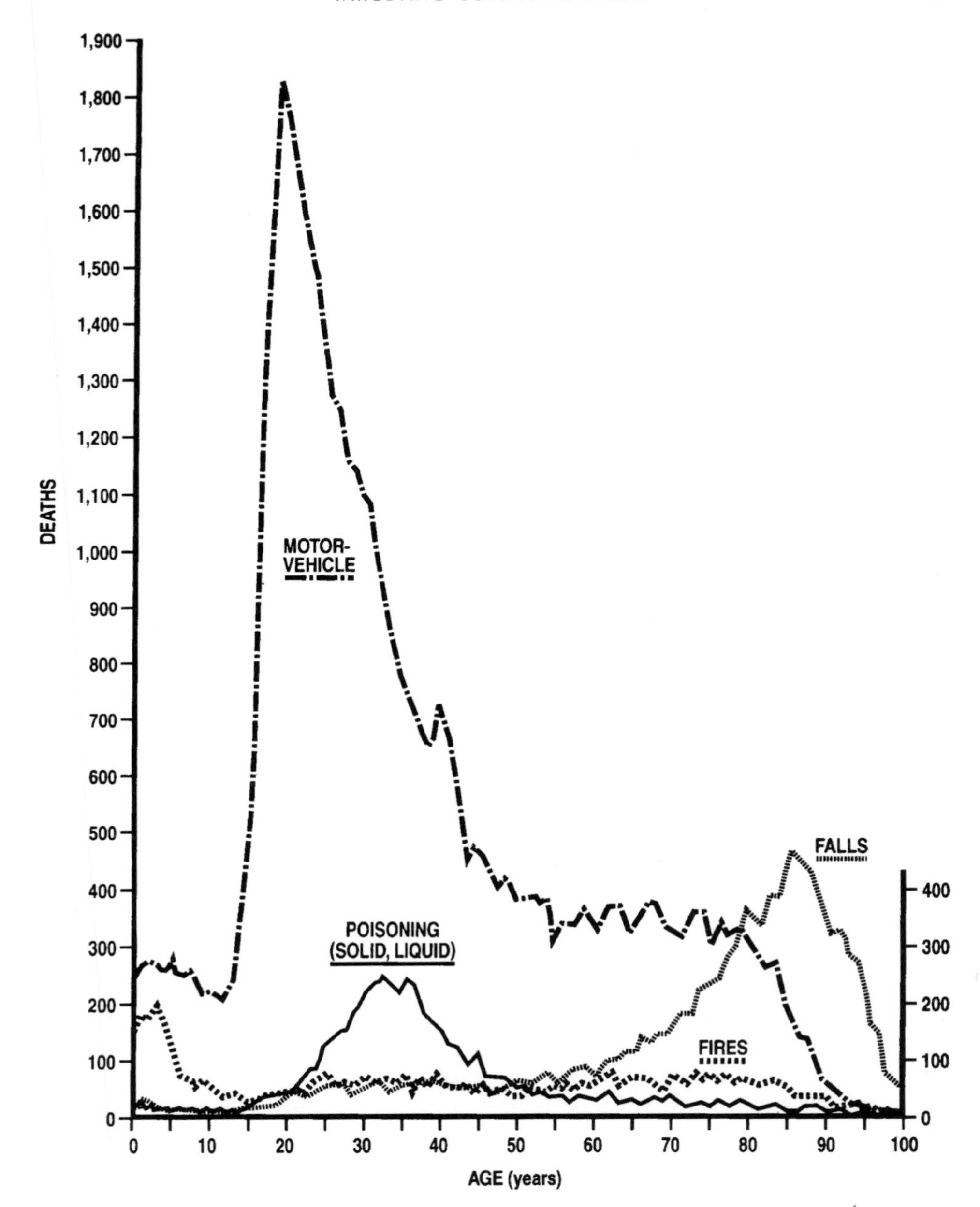

Figure 3.2 Causes of accidental death by age, 1988. 'Motor vehicle accidents were the leading cause of accidental death overall and the leading cause of accidental death from birth to age 78 in 1988' (National Safety Council, 1992, 9).

First definition of the target group: age and gender

A statistical information review reveals that the group formed by male drivers aged 18–24 is responsible for the highest number of casualty collisions per year of age. This group appears as particularly vulnerable when the number of casualty collisions per 1000 drivers is considered. (For this study, the 16–17-year-old group is left aside, although it is also highly vulnerable, as can be seen in the table. This is done because the objective is to plan a mass-media campaign, and most members of this latter group are reachable through the Alberta high-school system. Direct communication through the schools would be more efficient than mass media in this case.)

Table 3.1 Age and sex of drivers involved in casualty collisions: number, percentage of collisions and ratio per 1000 drivers

	Males			Females			Total*		
Age	N	%	1/1000	N	%	1/1000	N	%	1/1000
Under 16	312	1.4	20.3	126	0.6	11.0	438	2.0	16.3
16–17	812	3.6	27.7	443	2.0	17.8	1 260	5.6	23.2
18–19	1 101	4.9	29.4	534	2.4	16.5	1 639	7.3	23.5
20–24	2 453	11.0	23.1	1 035	4.6	10.7	3 496	15.6	17.2
25–34	4 264	19.0	14.9	2 013	9.0	7.8	6 286	28.1	11.6
35–44	2 664	11.9	11.6	1 438	6.4	7.2	4 107	18.3	9.5
45–54	1 450	6.5	11.0	722	3.2	6.5	2 174	9.7	8.9
55–64	955	4.3	9.8	371	1.7	5.1	1 328	5.9	7.8
65 +	812	3.6	9.7	291	1.3	5.2	1 104	4.9	8.0
Unspecified	143	0.6		58	0.3		556	2.5	
Total	14 966	66.8		7 031	31.5		22 388	99.9	

*Total includes drivers whose sex was unspecified in the collision report
Source: Alberta Transportation and Utilities, 1990

While men of 18–24 represent only 6.3 per cent of all Alberta drivers (119 266 out of 1 880 478), their involvement in casualty collisions represents 15.9 per cent of the total (3 554 out of 22 388). It could be argued that the definition of a high-risk drivers group requires consideration of miles driven rather than just the total number of collisions and the ratio per 1000 drivers; however, the indicators used are sufficient to select the chosen target group. The objective in this case is not to determine who is a better or worse driver, but which is a group that shows high rates both in total number and in the vulnerability of its members to the risk. Although drivers older than 80 are a higher risk per mile driven, a 50 per cent reduction of casualty collisions in this group would make no significant impact on the overall statistics, and would be ineffective in cost per subject reached through mass media. On the other hand, a 20 per cent reduction in the cohort chosen would be perceptible in the total number, and a change in driving attitude achieved in a young group would have a long-lasting effect as the group ages.

Even though young female drivers are also overrepresented in casualty collision statistics, males are responsible for more than twice as many collisions throughout almost all ages. Although both males and females show a high collision rate during their novice driver time, the rates of females tend to diminish, whereas those of males increase in the second driving year. This suggests that collisions among females have a high correlation with lack of driving experience, whereas other factors seem to affect collision involvement in males, revealing the possibility of attitudinal problems. This further reinforces the need to reduce the target group to the males, because it is likely that a campaign addressed to males will need to be different from one addressed to females.

Second definition of the target group: alcohol

After defining a particular target group, it is necessary to narrow down this choice further through the consideration of any dimension that affects the target group and establishes a division, relevant to either the problem or the people.

The objective of this campaign is to reach, within the chosen target group, the sober driver. It is not that drinking and driving is a dismissible problem, but there are already campaigns in place that have confronted this issue and show a positive impact: from 12 per cent of participation in casualty collisions of drivers who had been drinking in 1987, the number had gone down to 9.7 per cent in 1990, of whom only 4.4 per cent were legally impaired (Alberta Transportation and Utilities, 1990, 64).

It is time to confront the problem of the enormous percentage of drivers (90.3) who cause serious collisions when perfectly sober.

> In retrospect, it now appears that preoccupation with alcohol as a risk factor, particularly in the crash experience of young drivers, has proven a hindrance in identifying other factors that contribute to crashes involving young drivers, including those involving the use of alcoholic beverages. Many variables lie between alcohol as 'cause' and crash as 'effect'. A more complete understanding of young driver crashes (including those that involve alcohol) requires that we look beyond the presence of alcohol *per se* and consider the role of other social, psychological and behavioral factors that interact to determine crash risk. (Beirness and Simpson, 1987, 142).

Although we are not explicitly going to address the campaign at the sober driver, we are going to consider all the issues that could significantly contribute to serious collisions, but to the exclusion of alcohol.

3.2.4 Further definition of the problem: crash avoidance or post-crash safety?

The intention of this campaign is to affect the attitudes and behaviours of drivers so that they avoid collisions; we are concerned, that is, with crash avoidance. If we were concerned with post-crash safety, we would be discussing air bags, seat belts, rescue systems, and any other device or system that can prevent injury once the crash happens. This would require a totally different strategy. The reason that we are concerned with crash avoidance is that seat-belt use in Alberta is reasonably high, standing at 87 per cent since compulsory use was reinstated (it had fallen to 46 per cent while compulsory use was suspended because of a legal conflict). As in the case of drinking and driving, seat-belt use promotion, legislation and enforcement seem to be effective. Therefore, we prefer to address the issue of the reduction of the collisions themselves, which still keep creating a significant amount of suffering despite the existence of seat belts, air bags, ABS, rollbars and collapsible fronts.

Some authors suggest that larger cars provide better safety (Edmonston, 1990, 45–6). This single indicator, however, is insufficient. The USA Insurance Institute for Highway Safety studied the performance of 103 car models (1986–8) to determine a death-to-crashes ratio. They determined that the average was 2 deaths per 10 000 registered cars. Out of the 14 cars showing the highest ratios, 12 were marketed as sports cars, with the Corvette and the Camaro – not exactly small cars – at 5.2 and 4.9 deaths per 10 000 registered vehicles respectively. On the other hand, at the upper end of safety, the Cadillac

Fleetwood/DeVille and the Toyota Cressida (at different bulk) shared the same rate of 1.1, even though no perceptible difference in weight can be seen between the Corvette and the safest Volvo. Among the 10 safest cars of this study, the most expensive Volvos rated best at 0.6 deaths per 10 000 registered vehicles, several were station wagons, and one was a two-door (Cadillac Fleetwood/DeVille). None of the 10 best was marketed as a sports car. It is possibly true that Volvo's safety standards are high, but it is also true that a car buyer highly concerned with safety will most likely choose a Volvo over a Camaro. It becomes clear that it is not the weight of a car, its safety features or any other physical difference alone that determines the death-to-crashes ratio. Instead, the marketing strategy, the public image of the car and the psychological profile of the buyer are more important in determining the statistical direction. It is evident that an analysis of the attitudes of the drivers, and their corresponding actions, could help in focusing on factors that significantly contribute to and determine the occurrence of casualty crashes.

First conclusion

Given the above, the study will concentrate on the character profile, attitudes and actions of Alberta male drivers 18–24 years of age that make them vulnerable to becoming involved in injury crashes, without considering alcohol as a determining factor.

3.2.5 Objectives

The main objective of this study is to contribute to the reduction of traffic casualties in Alberta through the creation of information about drivers in the defined high-risk group.

Secondary objectives

1. Develop a profile of the target group.
2. Define reasons and motivations for the driving behaviour of the target group.
3. Develop criteria for a communications strategy aimed at reducing traffic injuries among the target group.
4. Recommend strategies for future action.

3.3 PROFILE OF THE TARGET GROUP

Every group is formed by different individuals, that is, the definition of the attitudinal and behavioural profile of a given group cannot be reduced to 'the group's average person's profile', because qualitative differences separate individual from individual. The cluster of attitudes and behaviours recognized in a given group, however, will most likely include some that rarely exist in other groups and exclude others that do. This helps to narrow the communications range, to eliminate some options, to emphasize others, and to define the spectrum of approaches that will form the strategy.

This is different from 'the shotgun approach', where the media and the public space are invaded by a great quantity and variety of communications. It is expensive and needs an enormously high budget to be effective. Instead, the approach proposed here is not unilinear, because it recognizes the plurality of profiles of the market mix, but it operates by recognizing the most salient characteristics of the market segment in question, and avoiding waste and the existence of messages that could be counterproductive to any segment of the audience addressed.

3.3.1 Method of study: focus groups

The purpose of a focus group is to generate qualitative information. The intention is to collect the emerging issues that surface in the discussions with the group. Normally, focus groups consist of a total population of 30 to 36 subjects divided into three groups, each group attending a meeting of about 90 minutes.

Usually, focus groups are combined with individual interviews and mail surveys to construct an audience's profile. In connection with this project, no personal interviews were developed: three focus group sessions and a literature review provided the initial material that was later complemented by a mail survey.

Selection of subjects

The process of targeting communications requires a careful definition of the target group. Although the various definitions described above were sufficient to establish a base, the selection of the actual members of the focus groups had to respond to those definitions, as well as to further narrowing criteria. Even though it had been decided that our group was the male driver 18–24 years of age, it was assumed that not all members of this group would be equally prone to risky driving. Actually, out of the 119 266 licensed drivers in Alberta (on 1 December 1991, according to the office of the Solicitor General), 3 554 were involved in casualty collisions in 1990, about 3 per cent of their age group. If we were attempting therefore to form the focus groups by just inviting subjects in the general universe of the target population, we would have a statistical chance of getting roughly one appropriate subject for every 35 members of a focus group. If we consider further the probability of response of different subjects and their willingness to discuss safe driving, the statistical probability stated above could be seen as highly optimistic.

Given the above, it was decided to form Focus Group 1 by inviting drivers who were repeat offenders, that is, drivers whose conviction record for moving violations had reached a level that required suspension and remedial driving courses. Focus Groups 2 and 3 were formed from a pool of drivers who had been involved in casualty collisions as drivers. It was decided further that these collisions should not have resulted in fatalities or severe injuries, to minimize the chance of psychological change that might occur in a reckless driver who survives a traumatic experience. This was done because the aim of the study was to develop a profile of drivers with attitudes that are likely to cause casualty collisions, and the sobering effect of a traumatic experience might render an originally appropriate subject as no longer representative of the intended target group.

Procedures for setting up the focus groups

Focus Group 1

To form this group, a list of 44 repeat offenders was provided by the office of the Alberta Solicitor General. Of these, 6 were not approached because they lived too far from the university where the meetings were going to be held. A letter of invitation was issued. This included a map showing how to reach the parking lot, a plan of the building with the location of the room, a reply form and a stamped, self-addressed envelope. A $30 honorarium and refreshments at the meeting were offered, as well as prepaid parking lot reservations or taxi rides. Out of the 38 letters sent, 8 were returned as 'moved' or 'wrong address', 20 were not answered, and 10 generated a positive response. The positive response to the letters, presumably received by the addressees, was therefore 33.3 per cent, a good level for this kind of invitation. All correspondence for the three groups was hand-signed by the principal researcher.

A few days after sending the letters, an attempt was made to telephone the subjects, but only for 23 out of the 30 could the telephone number be found. It was difficult to reach these subjects by telephone. They had moved, were out of town, did not answer, were unavailable, or indicated that on the meeting day they would be out of town or working night-shift. In one case, the subject came to the telephone after I answered a number of questions asked by his father. Some were concerned about whether or not this was 'a government thing' or whether or not it was compulsory. Generally, subjects displayed lower-than-average telephone communication skills.

Those who had agreed to participate entered their telephone numbers in the reply form. Most of them were contacted by telephone a couple of days before the meeting to confirm their participation. Nine of the 10 subjects who agreed to participate at the meeting requested parking reservations (I did not find out if any of them were under driving suspension at the time). A letter acknowledging the reception of the positive replies was sent to the subjects to confirm reception and to maintain frequent communications until the meeting date.

Because it was expected that some of the confirmed subjects would not show up for the meeting, announcements inviting subjects to the interview and to contact the principal investigator were posted in the Physical Education building, where many males of the appropriate age normally go for recreation. This was done because it was too late to contact additional subjects by letter, and because it was feared that the number of actual participants could drop too far below the desirable 10–12. Three subjects called and two were invited.

It snowed and was rather cold on the evening of the meeting (5 November at 7 pm). Directional signs were posted at various places in the building and a research assistant stood by, ready to guide the subjects. Six of the 10 subjects who had confirmed participation by letter and one of the two university students actually showed up. The group was therefore developed with seven subjects. The university student – who said that he had three demerit points in his driving record – provided somewhat different responses from the rest during the meeting, and had better communication skills. The transcript of the meeting used only the first names of the subjects to preserve anonymity. These first names were thereafter turned into randomly chosen letters of the alphabet for future reference. Only the principal researcher – not even the marketing consultant who moderated the group sessions – had access to the full names of the subjects. To facilitate communication throughout the meeting, nameplates bearing everyone's first name – including the moderator's and the researcher's – were placed in front of everybody.

At the beginning of the meeting, a questionnaire was given to the subjects, partly to test it before a large-scale mail survey was launched and partly to programme a task that could be performed individually while waiting for the rest of the group. Some participants filled it in there. A self-addressed, stamped envelope was given to the rest asking them to complete the questionnaire and return it. Only one actually returned it.

A letter of agreement to participate anonymously was signed by both the subjects and the principal researcher, along with receipts for the honoraria given to the subjects. This was also done for the following focus groups.

Focus groups 2 and 3

To form these groups, a list of 140 subjects whose names were recorded as having been involved as drivers in casualty collisions (as described above) was provided by Alberta Transportation through the office of the Alberta Solicitor General.

An amended letter of invitation, which included additional information, was mailed to 90 subjects. The change to the letter, which included a reference to the Solicitor General, was made in response to advice found in the literature concerning the writing of these letters, in the hope of obtaining a higher response rate (Rothe, 1987, 127). Just in case this second text was no better than the first, a repeat of the letter sent to Focus Group 1 (with only the dates changed) was sent to the remaining 50 subjects on the list. Out of the 90-person group, we received 14 positive responses (15.5 per cent), 5 negative and 1 excuse. Out of the 50-person group, we received 10 positive responses (20 per cent), 1 negative and 1 excuse.

Given the difficulties encountered in locating the subjects of Focus Group 1 by telephone, the new letter of acknowledgement sent to the positive respondents did not include a promise to call them, but offered to them the opportunity to call the principal researcher if they so wished.

Of the 13 subjects who promised to attend Focus Group 2 on 19 November, 12 showed up. Of the 11 subjects who promised to attend Focus Group 3 on 20 November, 9 showed up. All subjects had requested parking reservations. As had been done with Focus Group 1, Focus Groups 2 and 3 found directional signs posted in the building, and the same questionnaire was distributed to them as they entered the meeting room. Despite the higher verbal skills of the subjects, and their apparent higher willingness to cooperate in the project, the rate of completion and return of the questionnaire was equally low.

3.4 THE FOCUS GROUP SESSIONS

3.4.1 Driving and driver behaviour

Ten open-ended questions were posed to the group members by the moderator to elicit opinions from them. These questions were sometimes asked of each member at a time around the table; at other times they were posed to the group for whoever wanted to answer first. The questions centred around driving and driving behaviour, and were meant to generate responses from which one could extract the issues, opinions, perceptions and other attitude indicators of the target audience.

Rather than repeat all the individual answers, every question is presented here followed by the summary of responses prepared by Dr Finn (in italics), and by my own comments.

1. Tell us, what do you like most about driving?

The most common themes raised were freedom and independence, followed by convenience. Those citing freedom included some emphasizing the ability to get away from the implied restrictions of urban society and those who were freer to do what they wanted to in the city. Those citing convenience emphasized the greater flexibility provided when using a car compared with relying on Transit (public transportation). Other motivations for driving were for fun and to relax. While only a few participants acknowledged they enjoyed speed in response to this question, others suggested they agreed in their later comments. Nobody openly rejected these views or strongly suggested driving was purely a functional activity. (Finn, 1992, 4).

It took a while to prompt the subjects to mention things other than freedom and convenience as what they liked most about driving, as if they wished to project an image of maturity and good practical sense. Their first uttered reasons for driving (ability to go from A to B conveniently, not depending on the bus) do not seem to mesh well with comments that they made later, such as going at 130 km/h through a winding city road at 3 am, or taking the car for a drive in the country, or speeding to scare a girlfriend after an argument, or going for a ride with other members of the BMW Club, or making circles and the figure 20 in snowy parking lots.

It is evident that, although many actions of the members of this group can be dangerous, they would like to be seen by the institutions of society – represented by the university where the meetings were held – as people holding practical and no-nonsense values.

Communicational implications: It would seem useful to conceive messages that could appeal to their practical beliefs, however weak those could be. It is possible that if the messages dealt with those values, the target audience might pay attention to them. It is also possible that, if the target group were described in the messages as practical-minded, they might enjoy it, might gain some self-esteem (which they tend to lack) and might develop a good relation of trust with the campaign.

2. What kind of things do you dislike most about driving?

By far, the most common dislike was the behavior of other drivers. Most participants criticized other drivers for being undecisive, inattentive and poor drivers. A minority of participants disliked aspects of the driving environment; rush hour traffic, traffic lights and road conditions. Note that many of these dislikes could be seen as different constraints on the highly prized freedom discussed above. Very few participants mentioned anything related to the considerable costs of driving in financial, human or environmental terms. (Finn, 1992, 5).

As Finn says, the main problem perceived by the subjects was the bad drivers on the roads. This will be further discussed under questions 6, 7 and 8.

Consistent with specialized literature, these men regarded themselves as excellent drivers, despite the number of collisions, suspensions, demerits and non-reported violations that they have been involved in. It is almost paradoxical that they would overvalue their driving skill so much and that many of them, at the same time, would suffer from lack of self-esteem, possibly eroded by several social indicators, such as job opportunities and yearly income. It should be noted that the 1990 average income of 20–24-year-olds in Alberta (both sexes) was $14 897 (Canadian dollars) as opposed to $25 036 for the age group 25–34, possibly creating grounds for a sense of inferiority in the younger group *vis-à-vis* the older group (Statistics Canada, 1990, 113 and 115). It is also to be noted that, although casualty collisions spread evenly across socioeconomic levels, repeat offenders tend to show fewer years in school. This, in turn, tends to affect income and employability. Whereas Alberta males with up to grade 8 earned an average of $20 935 in 1990, those with secondary and some post-secondary studies averaged $27 721, and those with university degrees reached $48 458 (*ibid.*, 20).

As will be seen in response to question 4, the subjects were sensitive to financial indicators, sometimes to the exclusion of any other consideration. The combination of their interest in money with their lack of it, could easily result in insecurity and in low self-esteem. This, in turn, might create a wish to engage in a search for control over something else (a car?), in assertive/aggressive driving, and often in hatred for people, crowds, rush hour and other drivers, particularly if they are older, are women, or appear to have more money. In compensation, they fantasize about solitude, tranquillity and intimacy, including at most one other person in their pictures of pleasure (see later discussion of alcohol and car advertisements).

Communicational implications: It is possible that, given the above, this group would establish a good rapport with a campaign that would praise their driving skill, fitness and alertness, and that would ask them to use those abilities to protect others, as well as themselves, from the unpredictability, slowness and hesitancy of some drivers. An argument of this sort might push them to drive more attentively and, by offering them an important role to play in the traffic environment, might boost their self-esteem. The communications should remove money from the picture, and should dwell on the positive aspects of the target group, while avoiding authoritarian or admonitory tones that might make the subjects feel inferior (as they might have felt until recently in front of their parents and teachers and possibly now in front of their bosses).

3. How do you feel about driving when other people are with you in a car? Does that make any difference to how you feel about driving?

The group participants were relatively evenly divided between those who claimed their passengers made little or no difference to their driving, and those who indicated that they would drive more cautiously with certain types of passengers. The latter occurred for two types of passengers, namely their own children or adult authority figures such as parent or parent-in-law. A small number indicated they would still be inclined to drive more aggressively with some male friends or a girlfriend. Many more suggested elsewhere during the discussion, that this was a pattern of behavior they had previously engaged in, but had given up as they matured. (Finn, 1992, 6)

The general tendency faced with this question was to show as a first reaction a sense of self-control and detachment, similar to that in question 1. Initially, the subjects would say that they were unaffected by passengers in the car. In the course of the discussion, however, and in response to more specific questions, such as 'Does it make any difference if there is a baby in the car?' or 'How about when you drive with your in-laws?' Typically, three specific modes of driving were reported: driving alone, driving with subjects that called for careful driving (where the drivers felt either respectful or protective of their passengers) and driving with 'buddies' who enjoyed a bit of a thrill. In this last situation, the drivers reported being more daring than when driving alone. A comment from one subject, made when relating his collision (racing motorcycles with a group of friends down a winding city road in rush hour), strengthens further the point that these subjects tend to enact their risk-taking behaviour under peer pressure. The importance of peer opinion was also evident in the proceedings of the groups when graphic communications were discussed: groups 1 and 2 preferred one poster as the most effective (Figure 3.18, p. 90), and regarded another as dismissible (Figure 3.19, p. 91). A vocal member, who spoke first in group 3, praised the 'dismissible' poster of groups 1 and 2 as being the most effective, and everybody else in his group followed suit. It is difficult to believe that all subjects in groups 1 and 2 would independently agree on something, and that all members of group 3, belonging to the same target group, would agree on something else.

Unquestionably, the sense of social agreement is a strong component of these men's decision-making processes concerning value judgements and – possibly – driving behaviour.

Communicational implications: Given the above, and given the ability of one member of a group to lead the direction of opinion, it becomes clear that it might be advisable to use members of the target group as models in the communications. If anyone is going to appear offering any kind of advice, it should be a member of their group, with the clothing, the bearing and the environment they relate to and represent. The responses to this question showed that these men are highly sensitive to the people in their environment, either because they feel protective or respectful, or because they are sensitive to peer pressure. Given this, it seems that the communications should attempt to generate a change of behaviour in a leading segment of the target group, so that, through emulation, a ripple effect can be obtained across the entire cohort. This is a well-documented phenomenon: in mass media the opinion of group leaders is essential for the success of a campaign. It seems that the large number who could be defined as 'followers' tend to deposit more trust on people they know than on the impersonal message of the media. This adds another narrowing dimension for the definition of the target group of the campaign: that of having to emphasize its appeal towards opinion leaders.

4. What does the vehicle you drive say about you as a person?

Most of the participants raised relatively typical stereotypes about drivers of luxury, off road vehicles and sports cars. However, large numbers pointed out that circumstances would often mean that a person would not be driving the car which they would ideally like to be driving. A probable explanation for this was the fact that a significant proportion of the participants indicated that they had once owned a car (or motorcycle) which better matched their self image. But due to accidents and high insurance rates, many could no longer afford such a vehicle and were now using a vehicle they would prefer not to have to drive. (Finn, 1992, 7)

There are two main issues to point out in this case: one is that the subjects appeared to be only concerned with the possible financial dimension of the question, that is, more expensive or cheaper cars. It took a while to bring other implications of the question to the groups, such as a car defining the owner as 'outdoorsy', family-oriented, comfort-obsessed, sports-minded, safety-concerned, mechanically inclined, country & western lover, careless, counter-establishment, European-car fan or whatever myriad of possibilities for self-expression that a car image can offer. Although, when heavily prompted, they recognized the ability of cars to express a wide range of personalities and lifestyles, they always fell back on the car as a status symbol, a financial status symbol to be precise, and sometimes as a means to attract beautiful young women.

Most subjects did not own the expensive cars that they would have liked to, or that some of them had owned, and there are grounds to believe that, given the intensity of the feelings expressed, this affected their already discussed self-esteem. They seemed to see the vehicle not only as a demonstration of what one is but also as a realization of the self: no good vehicle, no good self.

The second point emerging from the responses to this question is that, instead of understanding the vehicle as a symbol of what one is, they understood it as a symbol of what one would like to be. When they said that the vehicle they were driving did not represent them because they cannot afford the insurance of an expensive sports car (because of their driving record and financial position), they failed to realize that their driving record and their financial position are indeed part of what they really are, and that, therefore, the car that they drove did represent them properly. This failure to perceive the evidence of reality is the same one that surfaces in their self-evaluation as drivers, where collisions and violations were not seen by them as indicators of poor driving.

Communicational implications: In connection to the first point raised, communications should try to separate the realization of the self from the purchase of the car. Too much emphasis on visible indicators generates too much tension for these young people, who seem to develop intense emotional investment in the kind of car that they drive, and in the way that they drive it. There would be a need to demonstrate that a vehicle, and the actions of drivers, are not the realization of the self. Specific communications addressed to the younger cohort might be geared at showing how easy it is to lose the privileges that they now enjoy (such as luxury cars or even the freedom to drive), and the need to look after them.

The inability of the young driver to be aware of reality is a difficult problem, but communications should, possibly through hard information, persuade them that, for instance, ABS will not stop their 120 km/h car in 10 metres, that their arms will not hold their bodies away from the steering wheel in a head-on collision where momentum multiplies one's weight 15 times, or that being smart is not driving dangerously, but driving safely. We will have to learn about the way in which the target group constructs reality to be able to affect that construction.

5. What makes a person a good driver?

By far the most common theme identified was for a driver to be attentive and alert so that they could anticipate problems and respond quickly enough to avoid them. A secondary theme which received general acceptance was courtesy. Only very few participants raised the issue of handling skills in different road conditions or sticking to the letter of the law. Several indicated that obeying the speed limits was not necessary to be a good driver – rather it was important to be travelling at the prevailing traffic speed. Although many participants indicated that they were now better drivers than they had been when teenagers, they tended to attribute it to maturity induced by greater reponsibilities and seeing accidents, rather than simply more driving experience. (Finn, 1992, 8)

6. What makes a person a bad driver?

Most participants identified behavioural characteristics of bad drivers such as being cautious, hesitant or intimidated by other traffic, and allowing other things such as car phones or alcohol to distract them from the driving task. A minority identified demographic groups such as women or old people as bad drivers, and there was only one strong objection to these suggestions, despite the fact that these groups have fewer accidents than young adult males. These groups were characterized as more cautious and slow to react to problem situations. Despite the fact that many of the respondents reported they had been in several accidents, nobody suggested that they themselves might be a bad driver now, although some did report they had been reckless in earlier years. (Finn, 1992, 9)

Questions 5 and 6 deserve a comment together, since they are strongly interrelated. Alertness was reported as the key dimension for good driving, as much as a lack thereof was for bad driving. The subjects considered themselves as being able to see and hear well, whereas people over 50 or anyone with grey hair was seen as seriously handicapped and potentially slow-witted. They seem to suffer from an excess of self-confidence when it comes to the evaluation of their ability to avoid collisions through their alertness. Given their record of collisions, it seems that they overestimate the use of alertness, and that this overestimation prompts them to drive beyond the limit of their ability to actually avoid a collision in an emergency. Although courtesy was mentioned, it was always in the context of complaining about the lack of it on the part of other drivers towards them, and seldom the other way round. They seem to have a self-centred perception of driving, failing to recognize the needs of the traffic flow as a demand for everybody to cooperate. They rather prefer to have everybody cooperating to let them advance faster.

Communicational implications: On the positive end, a communication strategy should take advantage of these drivers' perceptions of their own superiority, and show them the important role that they can play in the traffic environment. In this context, the value of personal courtesy to others should be emphasized. On the negative end, a communicational strategy should reduce the value of alertness and complement it with the value of a more realistic assessment of the skills of the driver and the possibilities of the car to avoid collisions in emergency situations. Although there is no doubt that alertness is a positive asset in driving, route planning and safe speed are likely to be more important dimensions in connection with both the efficiency of the traffic flow and risk avoidance. A communication strategy should emphasize the value of these criteria as pertaining to superior driving.

7. Have you ever been in an accident while driving, and if so, how would you describe what happened to generate the accident?

(This question was posed only to Focus Group 1. We used the word 'accident' throughout the sessions to match the subjects' terminology and to avoid intimidating them.)

In this group, all the participants who had received license suspensions indicated that they were themselves essentially responsible for their accidents. Only the participant recruited on campus blamed the road conditions or the other driver. (Finn, 1992, 13)

Some of the subjects joked about the number of collisions that they had had. This group appeared to believe that they were either out of luck or that driving naturally entailed the occurrence of collisions. Although they did not quite blame others, they described the collisions as inescapable. They reported that they themselves drove while suspended as much as their friends.

Their accounts of the collisions tended to be sketchy, showing limited verbal skills, except for the university student and another member. It would be reasonable to believe, however, that the subjects who agreed to come to the meetings might represent the top end of verbal communication skills in their cohort.

Communicational implications: There is a need to promote the notion that a key component of risk avoidance is the keeping of a good safety margin. 'We suggest that the key concept of traffic safety from the point of view of road-user behaviour is the *safety margin*' (Näätänen and Summala, 1976, 243). With this sentence, Näätänen and Summala close their 1976 book on road-user behaviour.

A communications campaign should recognize that these subjects either do not perceive risk, do not care about it, or are not aware of the possible consequences of risk-taking driving behaviour. The first alternative (deficient risk perception) requires education in both risk and hazard perception; a difficult challenge for mass communications that might have to be transferred to driving education and licensing practices. The second alternative (carelessness about risk) has to do with the value that the subjects assign to risk, and requires persuasive work directed at the development of the wish to avoid risk, and at making them see risk avoidance as a valuable dimension of driving and living. The third possible motivation of their attitude (ignorance of the possible consequences of risk) requires a combination of information and persuasion; however, this group will likely include a number of subjects who will not respond to a communications campaign. Having survived several crashes unharmed, some of them lack evidence of the ease with which a spinal cord can be damaged, or the legal consequences that might have to be confronted. If their experiences have not changed their approach to driving, it will be difficult to affect them through communications alone, and it will be necessary to act in association with legislation and enforcement, and with the cooperation of their employers and other significant people.

8. So what do you feel are the causes of accidents?

(Posed only to Focus Group 1.)

This more general question failed to achieve its real objective of probing further into the reasons for the respondents' accidents. However, the responses provided an indication that several members of this group remained quite cavalier about their driving record and generally denied they might need to change their driving behaviour. (Finn, 1992, 13)

Responses to this question demonstrated that the subjects, in the main, saw collisions as accidents, that is, as chance occurrences. One of them even said, 'That's why they are called accidents!' They failed to see the details that caused the collisions, and their responsibility regarding those details. When asked whether or not their collisions had helped them to improve their ability to avoid others in the future, they answered 'I never had the same accident twice'. Other typical responses were 'I got distracted' and 'I guess I was going too fast', all of them revealing a lack of responsibility for the occurrence. They seemed to consider that getting distracted or going too fast are normal dimensions of driving, as much as collisions, and therefore, if one chooses to drive, all this comes along. They seem to surround themselves with friends who have the same driving attitude and record, and therefore draw the conclusion that life is just like that.

Communicational implications: The fatalistic perception of the subjects about their collisions, and the way in which the word 'accident' removes responsibility from the driver, requires that a communication campaign concentrate on a change of terms, replacing 'accident' with collisions, crashes and driver error. It will be necessary also to coordinate action with journalists to remove the word 'accident' from the press, and other qualifiers currently used that remove responsibilities from the drivers, such as 'the weather was a factor' or 'the car went out of control'. The campaign should emphasize that cars are driven, that they are not independent, and that superior driving demands control, a term that the subjects of the target group hold in high esteem.

9. What do you feel was the cause of your reported accident?

(Posed only to Groups 2 and 3.)

Fewer participants in these two sessions suggested that they were themselves the cause of their accidents. Many attributed their accidents to mistakes or inattentiveness on the part of the other party to the incident. Common causes were poor judgement when turning left at an intersection, and inattention in rush hour traffic resulting in rear end collisions. (Finn, 1992, 13)

The issue raised by question 2 that 'the other drivers' are the worst aspect to deal with in traffic got reinforced here by the way in which the subjects reported their collisions. Inattentiveness (of the others) and road conditions were next to blame. From Alberta Transportation Statistics (1990 database), we know that 'unsafe speed' is the leading reported cause of casualty collisions (9.8 per cent), followed by 'ran off road' (8.8 per cent) and 'failed to observe traffic signal' (6.6 per cent). It is interesting to know that 'ran off road', basically a single-vehicle incident, is 8.8 per cent for this age group and 6.1 per cent for all drivers. It is important to note that running off the road is the leading driver action in fatal collisions, at 13.7 per cent, suggesting that a high percentage of deaths in this group might be attributed to this cause. Consistent with this, this group also shows a high percentage of single-vehicle collisions, standing at 27.2 per cent of all casualty collisions. This group also shows a slightly higher percentage in the second leading cause of death: driving left of centre (which causes 12.4 per cent of all fatal collisions). It appears as 1.8 per cent for this group, and as 1.6 per cent for all drivers involved in casualty collisions.

Table 3.2 Comparison between male drivers 18–24 and all drivers involved in casualty collisions expressed in percentages.

	Human action	Casualty collisions		Fatalities
		18–24	All drivers	All drivers
1	Unsafe speed	9.8	6.1	5.3
2	Ran off road	8.8	6.3	13.7
3	Follow too close	8.2	7.1	0.4
4	Failed to observe traffic signal	6.6	5.7	8.2
5	Left turn across path	4.0	4.9	2.8
6	Failed to yield right-of-way	3.9	4.5	5.2
7	Entered lane when unsafe	1.9	1.9	0.5
8	Driving left of centre	1.8	1.6	12.4
9	Improper passing	1.1	0.7	1.4
10	Improper turning	0.9	1.3	0.9
11	Backed unsafely	0.7	0.7	0.5
12	Failed to signal	0.1	0.1	–

Note: Percentages do not reach 100 because of the omission of collisions with 'unspecified causes' and 'driving properly', which normally occurs when only one driver performed an inappropriate action in a multiple-vehicle collision. The category 'driving properly' is included in the original publication because the table does not deal with drivers *causing* collisions but with drivers *involved* in collisions, including both the drivers who cause a collision and those who suffer as victims, thereby obscuring the issue of causality and misleading some to believe that, most of the time, collisions happen even though drivers are driving properly.
Source: Alberta Transportation and Utilities, Motor Transport Services Division, Planning and Statistics Branch, 1990 data.

The differences between the two groups are somewhat softened because the 'all drivers' group includes the men 18–24-year-old. Table 3.2 shows that both groups coincide in the leading six causes of casualty collisions, being significantly higher than the other six listed. The severity rating of the above causes, however, should result from an analysis of both casualties and fatalities. In this way, cause 8, although not high in all casualties, is the second cause of death and requires strong attention, whereas cause 3 normally results in low severity.

Single-vehicle casualty collisions numbered 968 for 18–24-year-old male drivers in 1990, or 6.7 per 1000 drivers, whereas 873 were suffered by 25–34-year-old male drivers, or 3.05 per 1000 drivers. This reinforces the idea that our target group tends to drive beyond their ability to control their vehicle twice as much as the immediately older age group. Excessive speed might account for other listed causes, such as running off the road (inability to recover control of the vehicle, possibly after an evasive manoeuvre, or taking too open a curve when turning into one highway from another), for driving left of centre (to have larger left and right margins, and therefore a higher sense of security), for failing to detect or act upon a traffic signal, or for failing to yield right-of-way. Seen this way, excessive speed can be defined as the cause of most reported causes, and would stand as a cause for collisions at 35.8 per cent for casualties and 44.8 per cent for fatalities. Considering that we have excluded the categories of 'driving safely' and 'unspecified causes', if we consider only the reported causes, excessive speed would represent 74.9 per cent of all causes for casualty collisions and 87.3 per cent of all causes for fatalities.

Communicational implications: When tailoring communications for the group under study, it is essential to analyze the leading causes of injuries and fatalities and base informational, educational and persuasive communications on the specific actions that result in the worst consequences. In driving school, in retraining courses and through mass media, there is a need to provide drivers with more specific strategies than messages such as 'be more alert'. Näätänen and Summala agree with the existence of this need when stating that 'general slogans and appeals used did not instruct drivers on specific methods for reducing accident risk, and therefore did not positively affect the drivers' behaviour' (1976, 9).

The statistics discussed above, the testimonials of the focus group subjects and the severity of the consequences of excessive speed suggest that speeding should be strongly addressed in a communications campaign, particularly when one considers that it is associated with several other causes. This does not mean that specific actions concurrent with speed should be ignored, but just the opposite: speed represents the thrill, the demonstration of power; it is the attractive feature of driving for many of these men and should be dealt with carefully. One possible way to deal with speed is by dealing with all the problems that it creates, without frontally attacking it. Another possibility is to enhance the desirability of a safety margin and, rather than attack speed, reduce its consequences by attaching a glamour value to safe driving.

10. What kind of good advice would you have for people?

The participants offered a wide range of suggestions at the end of the sessions. A common theme in the first session was learning from your accidents – perhaps from simulated experiences – instead of more real ones. The second group seemed more enthusiastic about the potential of educational programs, especially if they were targeted at junior high or even younger children. The final group tended to make suggestions for specific driving situations such as passing trucks on the highway and backing into parking spots. There was some reiteration of the need to use strong messages and vivid images when communicating about accidents. (Finn, 1992, 14)

The subjects also complained about lenient licensing practices and traffic laws in Alberta, and about the use of in-car telephones. Despite the fatalistic attitude that many of them showed towards their own collisions, all thought that something could be done so that younger people would not drive as recklessly as they had driven before close calls and dying friends sobered them up. Some acknowledged the difficulty of the enterprise and the apparent need of everybody to experience things personally.

Communicational implications: Three main message areas emerge from this question: (1) there is a need to introduce educational programmes as early as possible in the school system, well before driving age; (2) there is a need to develop instructional communications with specific information about appropriate and inappropriate actions; and (3) there is a need to produce persuasive communications to drive home the message that the risks are not worth it.

It could seem excessive to rely on the information emerging from the small sample of the focus groups to draft recommendations for a communicational strategy; however, according to the Paterson/Janzen Report generated for this project, the information obtained was comprehensive and consistent with the literature.

3.4.2 Reactions to graphic communications

To obtain information about the way in which the target group perceives printed communications, advertisements for alcohol, cars and traffic safety were shown to the subjects.

Alcohol advertisements

(Although we refer to the ads here by figure number, they were presented to the subjects identified by letter codes, to avoid any hierarchical suggestion in the presentation. The order of the figures was distributed randomly at each meeting.) The selection of the alcohol ads (Figures 3.3 to 3.7) was done to represent five different value sets.

The first task required of the group was to write down their ranking of the ads, using a form provided, from the one that they considered most effective in promoting the product to the one that they considered least effective. This was done in silence, and before a brief discussion of the reasons that everybody voted as they did.

Although there were many variations from person to person, some agreement existed in supporting Figure 3.4 as the best, followed by Figure 3.5.

The opening comments of the discussion were made from a distant and cool stance: 'I think [Figure 3.5] focuses on the product most. It's all black and white except for the product. So it really stands out', the participant said, disregarding the presence of the inviting woman (in the original ad, the female face is in black and white, only the product is in colour); or 'The Bacardi. It's showing you the product. So it really stands out.' These statements appear to be prompted by the same attitude that was aroused by question 1 ('Tell us, what do you like most about driving?'), where the subjects identified convenience as what they enjoyed most about driving, leaving aside all the thrills. The issue of relaxation, that is, changing from a description of the product to a desirable state, came up later several times: 'The Kahlúa one is good. It doesn't make you think of getting loaded or anything. Just relaxing.' '[The Kahlúa] it's relaxing . . . Not hyped and not pressured.'

Figure 3.3 Casual atmosphere, friendship and love, sunny outdoors, openness, middle-class, between 20 and 30 years old

Figure 3.4 Cosiness, interior setting, comfort, no people (reproduced with permission)

Figure 3.5 Eye contact with a woman, intimacy

Figure 3.6 Matter-of-fact product presentation

'[Kahlúa] I like the warmth . . . sort of reminds me of evenings after skiing. It's a lot more relaxed, and kind of laid back.'

Although it is impossible to speak more than about a few trends, it is interesting to note that, even though these men tend to be aggressive when driving, they showed a preference for the two ads that portray intimacy, warmth, solitude and relaxation, rather than the socialization-oriented ads. It was surprising during the focus group discussions, and particularly in Focus Group 1, to note how much these subjects were attracted to 'being

Figure 3.7 Upper-class, socializing, dressed-up party, nightlife

away from it all', either fishing and hunting, or just gazing at the horizon by themselves. They also longed to be 'in a comfort zone', as one of them put it in relation to Figure 3.5. It seems that, although they lead lives highly regulated by the impression that they want to create in others (partly through the cars that they own and the way that they drive), they feel tense, and they see solitude or the one-on-one relationship with a woman who would love them as the ideal places to be. In these places, they do not feel the gaze of others to whom they have to prove themselves.

Communicational implications: Although these men tend to drive aggressively and buy action and sports magazines, it is evident that not all the communications of a campaign addressed at them have to be designed along those lines. Instead, it appears to be useful to offer these men visual structures that would give them a sense of comfort, tranquillity and relaxation. It is evident, as said at the beginning, that any communication strategy will have to be supported by a variety of messages, ranging from the high-action ones to the peaceful, warm and secure ones. Because these men seem to be stressed by the roles that they believe they should play as drivers, perhaps one of the objectives of a campaign could be to reduce the social appeal of the 'supermacho', in favour of success with a more relaxed, unhyped, comfort-giving attitude.

Car advertisements

Ten advertisements with different thematic characteristics, car types and visual styles were chosen for this series.

The subjects were asked to fill in a form, indicating the three best ads and the three worst. Overwhelmingly, they chose Figures 3.17 and 3.10, as first and second respectively, leaving

Figure 3.15 for a clear worst. Figures 3.15 and 3.12 were the only ads that nobody chose among the three best, and Figure 3.17 was the only one never chosen among the three worst.

One subject chose Figure 3.9 because he liked the car, not the ad. When asked why he had chosen Figure 3.9 as the best advertisement, he proceeded to go through a long list of technical advantages of the Buick, practically without having seen the ad. This made evident the difficulty inherent in this kind of survey, and the importance of having a chance to discuss the choices of the subjects. People often seem to be so associated with the objects represented that they are unable to discuss the visual or textual characteristics of the presentation of the object. Some subjects, however, did make comments about the visual presentation. In many cases, their attention was caught by it. A couple of quotes illustrate the point: 'I like the Cadillac one because it's not a view you would see'. 'The Toyota Tercel one is good I think as well. Because it is zooming off, it's got some motion'.

From their comments, it is possible to draw several points: a great majority chose car ads that stand for a certain kind of rugged manliness, but not necessarily dirty ruggedness; a subject said: 'I just like the "Half man, half beast". I didn't like the picture. Dirty.' Sometimes the choice of adventure was watered down by a statement of utility: 'I guess the Nissan ad. It's got the practicality . . . it's got the fun aspect.' Nevertheless, freedom and power won over economy and practicality.

The fact that many of them chose the cars rather than the ads points to the strong connection that these people have with content over form. On the other hand, this gives an indication of the kinds of forms that they prefer, given the strong relationship between the styles of the vehicles and the styles of the ads. The subjects ended up preferring the same kind of car in two ads that represent, in different ways, fantasies of rugged adventure: the Nissan dwelling more on power, the Jeep more on prospect. ' "Only in a Jeep", and it brings up images of exactly what a Jeep is.' 'Just because it's wide open spaces, it looks really inviting. Looks like you'd rather be there than here right now.' 'I'd like to be there, looking at that.' Whereas nobody saw anything negative in the Jeep ad, none of the subjects saw anything positive in Figure 3.15. Several subjects made positive comments about the Cadillac ad (Figure 3.12) because of the presentation of road action; however, the car itself seems to have prevented the young men from choosing the ad as one of the best.

Figure 3.8 Practicality and economy, speed and wit

Figure 3.9 Romanticism, female appeal, softness, relaxation and affordable luxury

Figure 3.10 Ruggedness, macho animal, cut-and-dry, strength, beerhouse humour

Communicational implications: The strong connection that the subjects showed with the specific objects represented in the ads makes clear the need to pay special attention to the objects, places, people and, particularly, cars when choosing settings and elements for a communicational campaign. Although the clear primacy of Figure 3.17 seems to have been based on the appeal that openness, prospect, solitude and adventure have on these people, it cannot be denied that the Jeep itself played a strong part in the preference, making the portrayed freedom believable. This reinforces the notion that targeted communications should be based on an inventory of objects and places to which the audience can relate, as well as on the values that they appreciate. In this case, this small survey suggests that we can support the use of four-wheel drive, non-aggressive cars (we are not talking here about cars with extra-large tyres for all-terrain driving), and a sense of openness, solitude, freedom and, eventually, some elements of dynamism and speed.

Figure 3.11 Dependability, economy, comfort, order

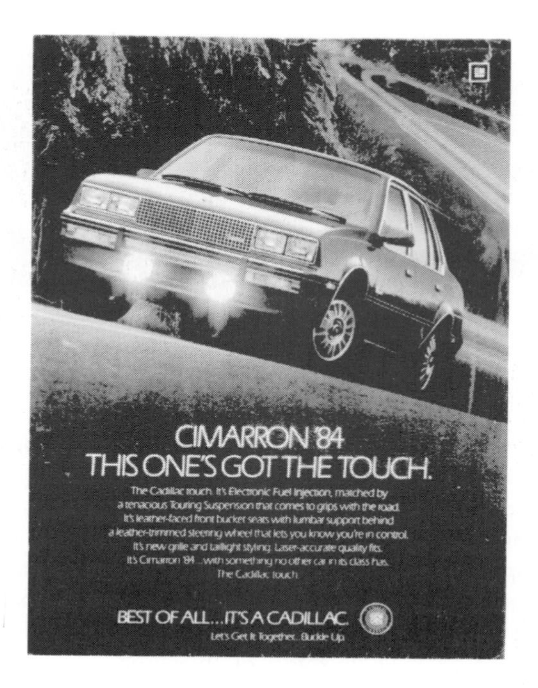

Figure 3.12 Action with luxury

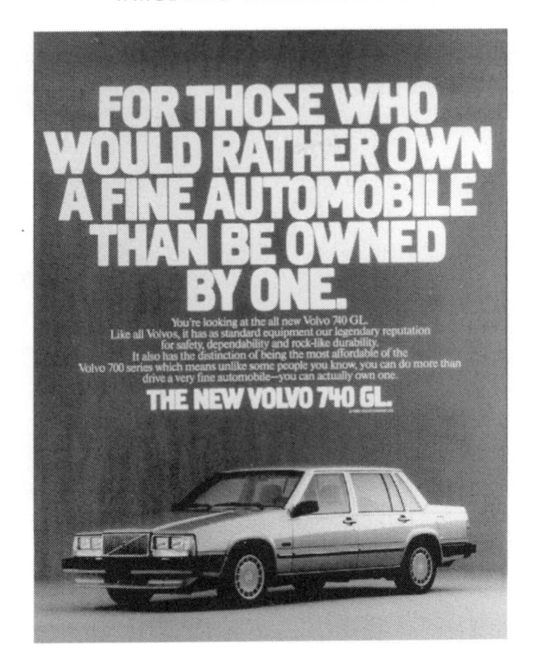

Figure 3.13 Safety, tradition and no-nonsense

Figure 3.14 Yuppie fantasy life, outdoors, free time

Economy cars and their sense of practicality could have some minor appeal, whereas family, traditional values and infantile imagery should be avoided. Witty titles would be appreciated, but long texts should be dismissed. The subjects seemed to prefer the layouts that presented a big and clear image and a clearly readable title over the others that had many elements, however neatly arranged those elements might have been.

Figure 3.15 Family vehicle, children's magic

Figure 3.16 Order, economy, careful buyer appeal, tidy, descriptive, unemotional

Safety posters

A third group of images presented to the subjects was formed by traffic safety posters produced by different agencies. The summary of responses to this material provided in the Finn Report reads as follows:

Figure 3.17 Freedom, openness, adventure, solitude

There was a strong consensus that strong messages with vivid illustrations were necessary if such posters or print advertisements were to have any useful effect. There was considerable support for strong images conveying the consequences of drinking and driving or accidents in general. In two of the groups, photographic realism was identified as more effective than a more representational drawn presentation of the consequences of accidents. When participants cited outside material that had made a strong impact on them, it was usually something which vividly presented the outcome or consequences of a driving related decision. Humour was not seen to be an appropriate communication approach. (Finn, 1992, 12)

Figure 3.18 was found to be the most powerful poster by Groups 1 and 2, whereas Group 3 supported Figure 3.19. Reactions to Figure 3.20 were somewhat divided, but more intense when negative than when positive. This poster was supported only by those who read it. Several subjects believed that if the poster had to be read to be understood it was not that good; some did not bother reading. The major criticism of Figure 3.18, which was brought up only in Group 3, was also connected to reading. After a vocal subject of the group started the discussion by saying 'if you don't read it, you don't know what it's about', most members of Group 3 said that the message was indirect and unclear. They argued that it could be about multiple sclerosis, paralysis or any related thing, but it was not clearly about driving. Groups 1 and 2, instead, found Figure 3.18 to be extremely sobering and effective. At the same time, however, they took a clearly different position about Figure 3.19, criticizing two aspects of it: a check-stop is not the worst thing that can happen to a driver, and a drawn illustration is not as convincing as a photograph. Group 3 praised Figure 3.19 as conveying all the sequence of events. Only one member of Group 2 found it quite forceful, but for a personal reason: 'I wouldn't like to see my Camaro like that,' he said.

Generally the subjects thought that Figure 3.21 would not have a sobering effect, but just the opposite. 'They are deadly serious. Bang! We are going to be saving your butt.'

'They look like an efficient group. I'd trust them.' It seemed that the poster encouraged risky driving by providing a safety net after the crash. This seems to reinforce the tendency of some motorcyclists, daring drivers and sportsmen, who suffer several broken bones but always get them fixed, to believe in some form of personal indestructibility. The poster appeared to be effectively promoting the image of the rescue team, but not necessarily safe driving.

Everybody agreed that Figure 3.22 was inappropriate. It made the whole thing look like fun.

Communicational implications: Despite the position of some researchers today who suggest that scare tactics do not work, all subjects in the three groups supported the idea that the stronger the message about the consequences of collisions, the better it is. They reported having sobered up their own driving only after being part of casualty crashes or after having friends die in traffic collisions when in high school. Although it might be true that positive messages are effective, it seems that this should not be extended to exclude messages that show the consequences of driving dangerously or carelessly. The recent success of the already mentioned Australian traffic safety campaign confirms the effectiveness of confronting people powerfully with the possible results of unsafe driving.

Texts should be brief, to the point, and preferably just reinforcing what the images

Figure 3.18 Wheelchair man

Figure 3.19 Check-stop, for carrying out Blood Alcohol Content tests

convey. If texts are essential for the comprehension of the images, they should be prominent enough to make their reading unavoidable. Images, however, should be dominant for two reasons: because of their immediacy and because the target audience might react negatively against admonitory or direct messages – images are normally less restrictive and more open to personal interpretation. This is in line with Barthes's ideas, who suggests that the text has a repressive value in relation to the meanings of images, and that we can see that the ideology and morality of a society are principally invested on this level (Barthes, 1985, 29).

Humour and cartoons should be avoided. There was a recognition that realism was needed, making photography an ideal medium for the creation of images. The apparent realism of photography makes it ideal as a medium when attempting to confront people with reality. For Barthes, photographs are messages without a code (*ibid.*, 5). I would argue that there are photographic codes; however, to the eyes of the general public, certain common codes disappear, rendering certain types of photograph as messages that apparently do not have codes, that is, they appear as not mediated, and are, in consequence, stronger.

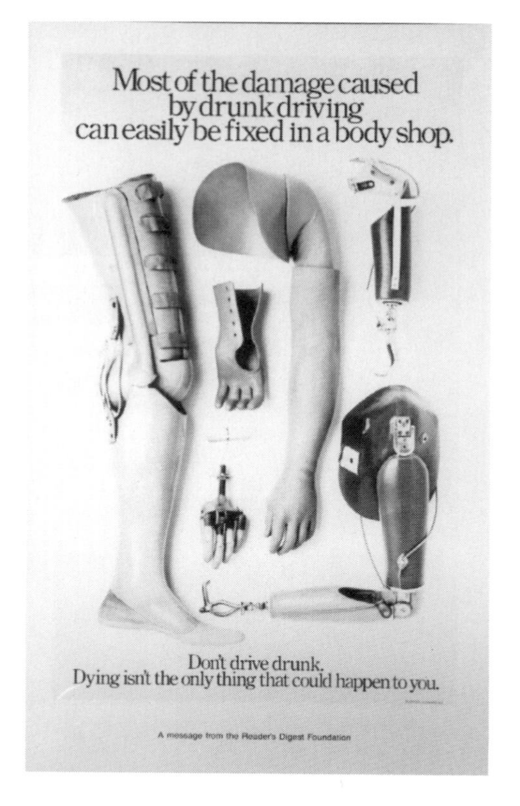

Figure 3.20 Artificial limbs

Figure 3.21 Rescue team

Figure 3.22 Cartoon

3.5 NARROWING DOWN THE TARGET GROUP

3.5.1 Personality profile of the 18–24-year-old high-risk Alberta male driver

The specialized literature, according to the Paterson/Janzen Report (1992), lists the usual personality traits of 18–24-year-old high-risk males as follows: daydreamer, aggressive, risk-taker, impatient, insecure, frustrated, angry, lacking self-discipline, lacking social conformity, becoming easily bored, being competitive, egocentric, having low self-esteem, getting easily distracted and being in a hurry, busy, active, nervous, and not easily accepting authority. He also tends to blame others, and drives to relieve stress and boredom, to get excitement, to experience power, independence and status, and to impress others. He tends to use driving as compensation for other unsatisfied needs (to look good, to feel sexy). On the negative side, he is proved to be fatigued or incapacitated by drink or drugs without noticing it. He tends to show inability to recognize hazardous conditions, lacks knowledge of the physical forces that affect a vehicle, and overestimates his skills to control the driving situation.

It is clear from the focus groups material, from the literature review and from common experience that driving is a complex social activity that far exceeds the function of travelling. In our society, it performs a highly symbolic function. It is embedded in many values that are strongly promoted by general education, mass media and the cultural environment, and is affected by social and economic problems that are well beyond the power of a communications campaign.

> [There seems to be a] lag between cultural values and social reality: schools, parents, mass communications media, etc., continue under the completely changed conditions of present-day society to foster values and ideas characteristic of pioneer days. American children are taught, just as their parents were, effectively and at a very early age that: a) aggressive competitiveness is useful from the individual standpoint and socially desirable; b) individual initiative and independence of action lead to success; c) one should seek maximal individual independence and control over one's environment; d) masculinity implies toughness, aggressiveness and the ability to withstand stress; e) challenge and excitement should be actively sought, and risk-taking is justified in meeting the challenge; f) status and other social rewards are more easily earned through individual effort than in cooperation with others. (Näätänen and Summala, 1976, 64–5)

It should be noted that the characteristics listed are by no means necessarily excessive in all men in this group. Contrary to this, if we look at the statistics, the majority show a good process of socialization, at least as far as driving is concerned.

3.5.2 A redefinition of the target group

As can be seen above, the development process of the target audience's profile begins by defining the age and gender of the target audience through statistical information on traffic collisions. Once the personality traits of the selected target group are defined, it becomes apparent that what we end up with is not a personality profile of the high-risk 18–24-year-old male driver, but the personality profile of a high-risk driver that appears more frequently among 18–24-year-old males but in other groups as well.

Given the particularity of images, which always have an inescapable precision, several campaign pieces will reach the discussed age group exclusively, but one can reasonably expect that the campaign will overflow those boundaries and reach individuals of similar character profile outside the age-and-sex group.

Communicational implications: It is evident from the above that one should only hold realistic expectations when it comes to changing the attitudes of those who have not found an appropriate balance between their freedom to drive the way that they wish and the safety of all concerned. There is, however, encouraging evidence that social groups are susceptible to changing their perceptions, attitudes and choices, as has been seen with drinking and driving, smoking in public places, and using a seat belt or a hockey helmet in Canada. Twenty-five years ago, manliness in hockey had no room for helmets, smoking was glamorous and drinking and driving was fun. The cultural meaning of these acts is quite different now. It should be learned from this that, where a change of a custom is pursued, a sustained effort is required. There is plenty of evidence in the literature that the 'safe-driving week' schemes do not have any impact, particularly after 'the week' is over.

3.6 RECOMMENDATIONS FOR A COMMUNICATION CAMPAIGN STRATEGY

1. *Main objective:* to reduce the number of traffic casualties.
2. *Subsidiary objective:* to aim at developing awareness, increasing understanding, changing attitudes and adopting new behaviours concerning safe driving.
3. *Specific objectives:*
   (a) On risk: to reduce the social value of risk-taking, and to teach risk-perception and risk-reduction skills.
   (b) On perception of reality: to increase the audience's ability to assess their driving skills properly, the car's ability to manoeuvre, and the severity of the risks faced.
   (c) On personal value vs. the value of things: to persuade the audience to value themselves independently from their car and their driving style.
   (d) On the value of responsibility: to promote the ideas that driving is a responsible act and that responsible driving has a high social value.
   (e) On competition vs. cooperation: to discourage competitiveness and aggressiveness in driving and to encourage cooperation toward a better traffic flow.
   (f) On the emotional component of driving: to reduce the emotional involvement that the target group invests in driving.
4. *Strategic materials:*
   (a) Instructional material concerning driving skills.
   (b) Informational material concerning the legal responsibilities involved in driving.
   (c) Educational material concerning driving judgement.
   (d) Departing from the values, knowledge and beliefs of the audience, the campaign should include persuasive material aimed at affecting driving attitudes and behaviour.
5. *Implementation methods:*
   (a) Decision-makers involved in the implementation of the campaign should be incorporated in the production team as early as possible to ensure their cooperation at the appropriate time.
   (b) The campaign must be sustained, intense and multichannel, using all media appropriate for the target audience. This is not to be seen as a 'shotgun approach', but as one that considers the most important emerging characteristics of the different members of the target group and the specific media with which they interact.
   (c) The campaign should seek favourable news coverage and gain the support of the entertainment industry for the promotion of its goals; it should aim at making traffic safety part of the social agenda, gaining support from both government and the private sector.
   (d) It should be complemented by a revision of legislation, and the enforcement and publicity thereof.
   (e) It should include a long-term plan divided into distinct phases, each having achievable, specific and measurable objectives.
   (f) Each piece of the campaign must concentrate on one point. Each conceptual point might require a number of specific pieces of communication.
6. *Campaign themes:*
   (a) The campaign should teach drivers to avoid situations and conditions that lead to collisions. It should centre on risk avoidance through strategic planning, rather than on risk evasion based on vehicle-handling skills.
   (b) It should teach specific points relating to safe driving, such as how to overtake

a truck on the highway, how to park safely and how to avoid the causes of collisions that are most common among the target audience, such as excessive speed, driving left of centre, running off road, failing to observe traffic signals, failing to yield right-of-way, getting distracted, or following too closely. It should centre on excessive speed, which is the cause of many reported causes.

(c) It should teach how to adapt driving style to the three environments that affect driving: the traffic environment, including light, weather, road and other road-users; the car environment, that is, people, pets, things or other sources of distraction in the car, and the car's performance; and the personal environment, that is, fatigue, drugs, illness, emotional problems, physical impairments, drinks, etc. It should foster the value of attention and the danger of distraction, such as paying attention to tape-deck operation, car telephone, sudden calls by others in the car, animals, map-reading, bill preparation for next delivery, and many other common distractions that negatively affect vehicle operation. It should put this in the context of planning one's driving and of maintaining an efficient and appropriate safety margin.

7. *Iconography for the campaign:*

(a) The campaign should use elements that belong to the environment, both real and desired, of the target audience. Specifically, the subject matter and its elements must be carefully conceived to include the appropriate cars (sports, pickup truck, four-wheel drive); the appropriate environments (lower middle-income motels, autoshops, sports stadiums, accessible winter and summer resorts, workout rooms); and the appropriate language (informal, direct, witty, colloquial – but not crude).

(b) It should use models in the age group of the target audience, performing appropriate tasks and representing the appropriate social classes; and other models representing people who are significant for the target audience (young women, parents, parents-in-law, small children); and the appropriate activities (sports, such as racketball, football, baseball, skiing, etc. – not golf or cricket). It should avoid authoritarian characters and the use of celebrities (people tend to remember the celebrity, but not the message; also, a celebrity who is an asset today can be a liability tomorrow) (Ogilvy, 1983, 25).

8. *Emotional appeal and persuasion:*

(a) The campaign should promote self-esteem among the target audience, treat them with respect and give them responsibilities based on the points where they consider themselves to be better than other people.

(b) It should reinforce the self-esteem of the audience by making them a role model for the younger and the older driver, by enhancing the positive values of their age, and by reducing their obsession with financial success and material possessions.

(c) It should use the audience's notion that they can handle a car better than anyone else (and teach them how to handle a car: what is possible and what is not).

(d) It should use the audience's notion that they are more alert than anyone else (and ask them to use that alertness to foresee other drivers' errors, to avoid collisions).

(e) It should use the notions of manliness and control, but change the actions that signify them (it is not advisable to label the notion of control as bad, but the actions that signify control have to change, and can be changed).

(f) It should capitalize on the target group's sensitivity to freedom, power, prospect, openness and adventure, and also on their alternative need for solitude, peace, comfort and intimacy.

(g) It should use the audience's notion that they know their city and its roads (and teach them how to use that knowledge better and how to deal with less knowledgeable drivers).

(h) It should use the audience's notion that they have better sight and hearing than anyone else (and teach them how to take advantage of that for the purpose of risk detection and assessment).

(i) It should promote the benefits of being a good driver and redefine what being a good driver really is.

(j) It should make safe driving more socially and personally attractive than risk-taking driving, building on existing motives, needs and values of the target group.

(k) It should associate safe driving with a positive reference group of the target audience.

(l) It should highlight short- and long-term benefits of safe driving.

(m) It should show the indirect benefits of safe driving in social and financial terms.

(n) It should inform about the physical, emotional, financial and legal consequences of traffic collisions, and of being at fault therein.

(o) It should use both positive and negative messages, tailored to fit the varied emotional makeup of the audience.

(p) It should promote interpersonal communication among the members of the target group, and between them and others, to reduce the use of driving as a compensation for unsatisfactory communication at other levels.

9. *Communicational tone:*

(a) The campaign must be non-judgemental, non-authoritarian and non-admonitory. It should seek partnership with the audience.

(b) It should not use humour or cartoons.

10. *Visual symbolism and structure, and verbal language:*

(a) The visual structure of the messages should connote the values referred to in recommendation 8(f), dwelling on the dynamic and expansive notions of freedom, power, prospect, openness and adventure, and alternatively on the calming concepts of solitude, peace, comfort and intimacy.

(b) The visual style should relate to the visual environment of the audience, as created by the television and the printed media that they generally choose, but excluding those styles that, through their sheer structure, promote negative attitudes towards society.

(c) Excessive ambiguity and complexity should be avoided, in favour of clarity, simplicity and directness.

(d) Preferably, images should be photographic, and should carry the burden of the messages, allowing texts to be a reinforcement, particularly in print media, and not dependent on text for clarity or appeal. Brief texts, titles and slogans in plain English are advisable, however, to aid memorization and conceptualization.

(e) The word 'accident' should be avoided, using rather the words 'crash', 'collision', 'casualty', 'rollover', or any other appropriate descriptor. Language should be controlled with the intention of avoiding any term that would remove responsibility from the driver as an active agent in the generation of collisions. In many cases, the terminology used by Alberta Transportation should be adopted (casualty collision, injury collision, fatal collision, or collision).

Cooperation with the entertainment industry and the journalistic media should be sought to eradicate terms that remove responsibility from the drivers.

11. *Implementation and evaluation:*
    (a) Elements and component pieces of the campaign should be pretested with the target audience before final production. This can be done through focus groups, individual interviews and surveys to get the target group's opinions. Although this can only provide indicators of acceptance or rejection and not of attitude change, it will help to increase the probability for actually reaching the audience.
    (b) The campaign should be launched in a relatively controllable setting, such as a town of around 100 000 people, to evaluate its performance and that of its components before mass implementation. The selected city has to be small enough that costs are manageable, but large enough that no exceptional individual case would substantially affect the statistics.
    (c) The testing period should be long enough (four to six months) to avoid distortions caused by exceptional events.
    (d) All usual collision statistics dimensions should be measured after the intervention. Although the final evaluation of campaign success will be given by the number of casualty collisions, specific evaluation instruments should be created once the campaign and its deployment are designed to assess the relative value of each component, so as to maximize the investment efficiency.
    (e) A non-intervention, similar town should be selected for control, and comparisons should be made between both towns and general Alberta statistics. This should be done for both the particular test period and for the last five years to assess the change obtained in the target city compared with the changes operating over the broader environment.
12. *Beyond communications; licensing and driver training:*
    (a) The campaign should be supported by the development of more demanding licensing examination requirements, and ideally by a system of graduated licensing.
    (b) Safety education should form part of general education from early on in the school system, in an attempt to increase the general concern for safety.
    (c) Driver education should become compulsory, but without encouraging early driving.
    (d) Driver education should add to its current curriculum persuasive elements aimed at instilling drivers with an appropriate attitude, beyond the technicalities on which they concentrate today.
    (e) Existing compulsory retraining courses for repeat offenders will have to be reassessed, along with legislation and enforcement related to dangerous use of a motor vehicle.

3.7 FUTURE ACTION

To complete the campaign plan, it will be necessary to develop a survey to find out media preferences of the target group. Further, it will be necessary to develop a series of campaign themes and test prototypes as indicated above. Finally, a campaign concept, a media strategy and a production and deployment budget will form the key elements for actual implementation. For the campaign to be effective, the budget would have to at least match the level required to launch a new consumer product in the region.

3.8 THE CAMPAIGN CONCEPT: FOCUS AND CHOICES

The high number of campaign themes will have to be subordinated under a few main concerns to gain a presence in the public's mind. It should be remembered that any mass-media effort has to compete in a field where 'advertisers spend over $50 billion annually in the United States and political candidates a half-billion more' (McGuire, 1985, 233). A strong identity will have to be created, so that each campaign component will reinforce the others.

3.8.1 Key psychological change to be pursued: reduce emotional involvement in driving

It is the symbolic aspect of driving that appears to be at the centre of the problem that we want to confront, that is, driving as communication. The campaign should reduce the social importance of driving and minimize the magical power that it seems to have for some.

Because this is a long-term goal, in the short term we should intend to shift the meaning of some of the emotional components of driving that appear as being extremely popular among the target audience. *Control* and *power* are two of the major key motivations that recur in the statements of these drivers. Control should be redefined to mean farsighted risk-avoidance planning rather than successful handling in risk situations. Power, instead of being based on hp, cc and speed, should be based on the power to be trusted as a driver – on personal power instead of engine power. As said above, we know that driving is a complex social act, and many driving behaviours have to do with either the actual presence of others, or with an internalized 'other' for whom the driver performs. The structure of this 'other' is highly dependent on the perception that the driver has of societal values. The campaign should work on both those perceptions and those values.

Whereas interviews with teenage boys showed that they saw sports cars and speeding as essential to attract girls, interviews with girls revealed that they preferred calm, safe drivers. It seems that lack of communication between the sexes makes boys engage in behaviours that they wrongly assume appeal to girls. This provides an interesting entry for the promotion of a change of attitude. If what the boys want is girls and not speed, there is room for bringing to light the girls' feelings and offering a strong motivation for change. We know, however, that boys like speed as well, and this will require a specific strategy.

3.8.2 Key operational change to be pursued: reduce speed (increase safety margin)

Obviously, if cars went at 10 km/h we would not have many collisions, and fatal ones would be rare. Without going to extremes, it is clear that excessive speed is the main factor in many reported causes of collisions. There is a need to remove the social prestige of speed (difficult, with big prizes for the fastest drivers in car races). A cool analysis of a trip in town will demonstrate that there is not that much time to be gained by going fast. Traffic lights, curves, other road-users and many other factors create delays. Reaching a final destination significantly faster in a 20-minute drive is extremely difficult to accomplish. Excessive speed is not the consequence of a practical mind, but more the result of a social value, the thrill that comes with information overload, and the sense of excitement that it gives to some people who like to feel that they live literally, as well as figuratively, 'in the fast lane'.

The main problem created by speed is the frequency with which drivers overestimate their own driving skills and their car's ability to challenge the laws of physics. Reduction of speed, and the related increase of the safety margin, are the two main operational changes to try to effect.

3.8.3 Planning vs. handling

Popular culture has it that a good driver is one who can come out well from an emergency situation. There is a need to change that, defining a good driver as one who avoids emergency situations altogether. Forward planning, road knowledge, foreseeing the errors of other drivers, and being aware of traffic ambiguity, should gain value as defining good driving, as opposed to the senseless attempt to control a car beyond physical laws and human abilities (as the entertainment industry persists in showing as a possibility).

3.8.4 Understanding traffic vs. understanding driving

Driving schools should be renamed traffic schools. There is a need to remove the self-centredness that pervades the notion of driving. The total picture of traffic flow should take precedence over the individual driver's perception. There is a need to promote the idea of driving as partnership with all other road users, rather than as a 'one-against-all' situation.

3.9 VISUALIZING IDEAS

3.9.1 Sorting the requirements

The first task to face after the development of the recommendations is their grouping and articulation. This must be done to use those recommendations as operational concerns that inform the communications conception. At the beginning, the recommendations were developed as qualitative information emerging from selective contacts with the audience and quantitative information generated by surveys, contacts with experts and literature review. Initially, all information was treated as qualitative, that is, looked at and noted insofar as it represented an original concern, regardless of the frequency with which that concern appeared. This was done to define the spectrum of audience perceptions, and to avoid falling into the reductionist trap of selecting only those concerns that appear in a majority of subjects. A large group of people (as is this case) does not necessarily include a majority of subjects who share specific traits and concerns, but it is formed by a large number of 'minorities' who suffer among them, for different reasons, a similar problem, in our case, a vulnerability to collisions. The same age and gender cohort is overrepresented in statistics as both causer and victim of traffic collisions. It would be reductive to believe that there are only two subgroups in this group (victims and causers), because there are quite possibly different reasons that people in both subgroups are vulnerable. All issues raised by the information-gathering process were considered, without discarding infrequent or isolated concerns, although frequencies and patterns were indeed sought to establish priorities and to define central and secondary elements. This information came in streams of unorganized units. To organize the recommendations in clusters and hierarchies, every individual item was written on a small card. Thereafter, groups and headings were created, reducing redundance and obtaining a more global vision of the different decision levels to be confronted.

At this point, the designer must restate the recommendations in his or her own way and introduce more vision-related criteria and definitions.

3.9.2 Test material

The preparation of a campaign requires the transformation of the recommendations in concrete messages. It is advisable to prepare test material creating specific pieces that represent clusters of those concerns in different degrees of importance.

To develop visual possibilities, the author, working with the Alberta Motor Association, organized a project with design schools in North America, through which more than 300 proposals were created. Figures 3.23 to 3.30 show examples of the entries that were recognized among the best by an expert group. Subsequently, the material has been used to study reactions of different groups.

3.9.3 Testing

After preparing a series of images and texts judged to represent the most important points to be addressed, a survey should be conducted with the target audience to fine-tune the approach before beginning expensive production and implementation. Iterative testing

Figure 3.23 Bryce Beresh, Nova Scotia College of Art and Design

Figure 3.24 Nurhayat Elturan, Indiana University

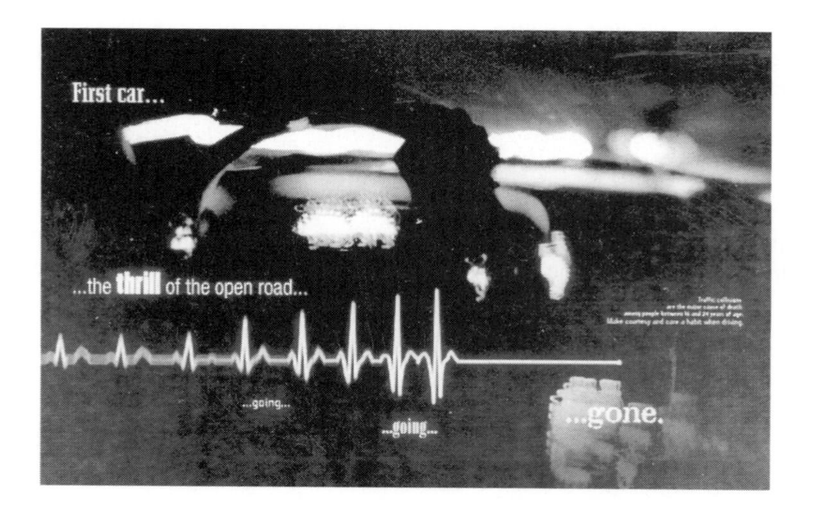

Figure 3.25 Jerry Demchuk, University of Alberta

should be conducted until the responses obtained indicate with confidence that the content and tone of the communications are promising.

Because the final measure for this kind of campaign is the number of collisions, and because this cannot be measured until final implementation takes place, the purpose of the tests is the exploration of general and specific negative and positive reactions from the target audience. The following conditions should be checked:

1. The themes used are appropriate.
2. The image content meets audience requirements.

Figure 3.26 Ben Guyer, Nova Scotia College of Art and Design

Figure 3.27 Sungmi Oh, Carnegie Mellon University

3. The visual production uses an appropriate level of resource.
4. The visual style is directed at the target group.
5. The texts are adapted in length and structure to each medium used.
6. The verbal style of the texts is carefully tailored for the target audience.

Given the impossibility of ascertaining campaign effectiveness before actual implementation, deployment of the final campaign in a small town is advisable before substantial investment in media space takes place with a view to affecting a large region.

Figure 3.28 Selene Yuen, University of Alberta

Figure 3.29 Stacey Shapiro, Carnegie Mellon University

Figure 3.30 Barbara Tejada, Carnegie Mellon University

CHAPTER FOUR

Case histories

4.1 INTRODUCTION

The cases reported in this section stand as examples of effective visual communication design within a socially oriented framework. There are many examples of effective communication design in the advertising field for consumer goods and services, but there is a need to orient that experience in the direction of effective communications related to social problems and to the provision of necessary services to people.

It is also necessary to demonstrate that public service design work should not be done as a charity, because good work pays for itself several times over, and requires serious investment. If a job is done as a charity, it is normally not done well, and therefore it is better not to do it at all. To generate attitude changes and to solve difficult cognitive problems requires excellent intellectual resources, good research and testing, and high implementation budgets for media space and time, things that cannot normally be done as charity. It is also necessary to report on successful socially oriented communication projects so as to demonstrate that socially oriented design work is not a utopian impossibility but a possibility that has been successfully realized in several cases.

4.2 AUSTRALIA'S TRANSPORT ACCIDENT COMMISSION CAMPAIGN

Editorial note: The text that follows is based on, and extensively quotes, the report published as 'Road Safety: the Transport Accident Commission Campaign' in *Effective Advertising, Casebook of the AFA Advertising Effectiveness Awards 1990*, Advertising Federation of Australia, Sydney, pp. 191–205. By permission received from Grey Advertising and the AFA.

In November 1989, Grey Advertising was approached by the Transport Accident Commission (TAC) of the State of Victoria, Australia (third-party liability insurer) to develop a campaign with an annual budget of six million Australian dollars. The objectives of the campaign were to reduce the road toll dramatically in Victoria by means of compelling drivers emotionally to change their attitudes and behaviours and by communicating two new police control measures: the introduction of speed cameras and alcohol test buses.

Grey Advertising, under the direction of Greg Harper, working with Leon L'Huillier, executive chairman of the TAC, developed and implemented a campaign that, in a little more than twelve months, resulted in the following:

- a 30 per cent reduction of the road toll, equivalent to 230 lives saved, compared to the previous year, matching the level of 1955
- $361 million in savings for the community
- $118 million in savings for the Transport Accident Commission in insurance payments.

The campaign enjoyed a 92 per cent unaided advertising recall, and 98 per cent of those surveyed were in favour of it.

4.2.1 Conceiving the campaign

The difficulty of the campaign was complicated by the existence of other traffic safety messages issued randomly by several agencies. The new campaign was conceived as if traffic safety were a consumer product, generating an identity for it to aid recognition, recall and loyalty. The conception of the campaign was heavily based on extensive qualitative research conducted by Brian Sweeney & Associates, whose initial research identified some 'dos' and 'don'ts':

- Don't concentrate solely on twisted metal.
- Don't overdo the statistics.
- Don't lecture.
- Don't suggest that he or she cannot have a drink.
- Do be as shocking as you like.
- Do be as emotional as possible.
- Do ensure that they come away from any communication thinking 'that could happen to me'.
- Do emphasize the 'drink/drive–speed/accident' link.

According to Grey Advertising, they were encouraged by the client to 'outrage, appall and upset' the community into confronting the issue head-on. The campaign had to prove effective in qualitative research before any investment in media was committed.

In developing the communication strategy, they based their thinking on important marketing principles, such as *launch strong, new product (or brand) names*. It was decided to concentrate on two 'names': 'drink/drive' and 'speed'. These generated two slogans: 'If you drink, then drive, you're a bloody idiot', and 'Don't fool yourself, speed kills'.

The second marketing principle followed was *to develop a single, powerful proposition*.

> Research indicated that the key to developing an effective proposition was to have the courage to confront the community in a way that would strike deep at their core emotions, personal fears and feelings of vulnerability. Prior to the campaign going to air in pre-Christmas 1989, the greatest single deterrent (against dangerous driving) was the rational fear of losing your license. Only six weeks later, the emotional traumatic 'life sentence' of finding yourself responsible for the death or disablement of another human being, had become a far more powerful and persuasive deterrent.

The third principle applied was *to segment the target market*. This meant recognizing the need to develop a wide range of different ads to reach the diverse groups of drivers.

The fourth principle was *to develop brand loyalty*. It was critical that the public believe the campaign to be relevant to their immediate needs, values and lifestyle. The messages were to be totally credible as well. Both in their general content, as in their most minute details, they were to avoid creating the slightest suspicion of being staged.

> It was critically important that people saw the ads and spontaneously thought: 'that could be me . . . that could so easily happen to me . . . !!' They had to believe what they saw was real. Real accidents, real emotions, real people suffering real pain and trauma. This required real people who have to deal with the horror, trauma and human waste, every day of their professional lives. People like Sister Karen Warneke, Charge Nurse responsible for Intensive Care at Royal Melbourne Hospital; Mobile Intensive Care Ambulance Officer Paul Thresher and Doctor Michael Johnson, a surgeon at St Vincent's Hospital . . . Realism required shooting without structured dialogue or staged scenes, and trusting a cameraman to get it all on film without ever seeming to intrude in the action.

This realism also meant misleading members of the public who witnessed the scenes and believed that they were real. The believability of the films produced suggest that it was worth it. This campaign succeeded in having an impact where many others had failed.

The fifth principle was *to merchandise and promote the product (message)*. 'Our aim was to build the kind of market presence for the "drink/drive" and "speed" products (brand names), that high profile consumer brands like Coca-Cola, Mars and McDonald's strive to achieve.' The campaign used extensive outdoor promotion, including strong, visible promotional support at major sporting events.

The major components of the campaign were five 60-second television films: *Girlfriend* (drink/drive), *Beach Road* (speed), *Speed Cameras* (speed), *Tracey* (speed) and *Booze Buses* (drink/drive). 'Research clearly demonstrated that a 60 second length was required to break through the surrounding programme and advertising clutter, and allow time for the viewer to become completely involved in the emotion, action and proposition on screen. The media budget for 1990 was (Australian dollars) $5.5 million. Use of media was: 70% television; 14% press; 7% radio; 5% outdoor; 2% Sky Channel; 2% cinema.'

According to the Australian authorities, the films could only be shown at restricted times to avoid affecting children. This meant weekdays from 8.30 am to 4 pm, and every day after 7.30 pm, exercising judgement to avoid programmes that attracted children within those times. Free community service airtime received by the campaign amounted to more than $300,000. The first film went on air on 10 December 1989. 'By the end of January '90, the road toll had dropped 37% below the figure for the same month in 1989.'

Possibly one of the most successful aspects of the campaign was its penetration in the mass media as part of the community agenda. Between 12 December 1989 and the end of January 1990, the campaign or its content was the subject of the following: 17 major press articles; 15 major regional press articles; 44 times on metro radio news; 16 talkback radio debates; 2 major features on television current affairs; 4 times on metro television news; and 3 times on regional television news. 'This trend continued throughout 1990, with 869 recorded incidents of the "drink/drive" and "speed" message impacting in the news media, and in the community itself.'

4.2.2 Campaign results

Public statistics comparing 1989 with 1990 show the following figures: 230 fewer fatalities (–30 per cent); 811 fewer permanent or serious injuries (–20 per cent); 6 187 fewer minor injuries (–19 per cent); 6 716 fewer collisions (–14 per cent).

Speed

Before the introduction of the campaign and the speed cameras, random checks had shown that 23 per cent of cars were exceeding the speed limit. By June 1990, similar checks showed that only 7 per cent did it. In addition, there had been a 30 per cent fall in the number of vehicles exceeding the limit by more than 20 km/h in 100 km/h zones, standing at less than 3 per cent. A significant reduction (28 per cent) of 'run off the road' crashes further confirmed the general reduction of excessive speed.

Drink/drive

The percentage of drivers or riders killed in 1990 who had a blood alcohol content (BAC) of over 0.05 per cent was reduced from 36 to 22 per cent. Random police checks also showed a related reduction in the number of drivers found with a BAC in excess of 0.05 per cent: from one in every 300 to one in every 500.

4.2.3 Cost-effectiveness

The cost of traffic crashes in Victoria in 1989 was estimated to exceed $1.5 billion. The office of Steadman & Ryan, working for the Bureau of Transport Economics of Canberra, estimated that in 1990 the community saved $361 million because of the reduction in road crashes. This includes losses of income to the victim and family, and the general impact of an incident in creating expenditures in hospital, medical services, rehabilitation, legal and court costs, insurance administration, incident investigation, losses to third parties, vehicle damage and traffic delays.

The savings to the third-party liability insurer who launched the campaign and the related control measures was 17 per cent, equivalent to $118 million, more than 20 times the cost of the communications campaign.

4.2.4 The overall strategy

The road safety campaign was a combination of police presence, the introduction of new equipment and a communications campaign. One cannot attribute to any factor in isolation the results obtained in Victoria; however, the results started showing strongly from the beginning of the campaign, before the measures came into full effect. The first two BAC-measuring buses ('booze buses') were introduced on 10 April 1990, and the total fleet was completed in the last quarter of the year. The first two speed cameras entered operation in December 1989, and an additional 10 were introduced later in 1990.

Since other Australian states incorporated the measures but did not use the communications campaign, one can isolate the effect of the campaign from the combined effect of the measures. It can be seen that whereas the overall impact of the programme in Victoria resulted in a reduction of fatalities of 30 per cent, the rest of the country, excluding Victoria, reached a reduction of only 11.6 per cent. By March 1992, fatalities in Victoria had fallen by 42 per cent, compared to 30 per cent in New South Wales, 18 per cent in Western Australia, 17 per cent in Tasmania, 15 per cent in South Australia and 9 per cent in Queensland. At that point, neither Queensland nor Tasmania had yet introduced speed cameras.

In 1989, the number of traffic fatalities in Victoria was 17.9 per 100 000 population, the highest in Australia. By the end of March 1992, it had fallen to 10.1, becoming the lowest in the country.

4.2.5 Attitudes and behaviours

According to a report published in *The Globe and Mail* in Canada (26 April 1993, p. B1), the interruption of the campaign in Victoria (possibly because of lack of budget) during May and June 1990 resulted in a jump in fatalities from about 42 to 55. Although no result is visible in an interruption during October and November of the same year, an interruption in March 1991 shot fatalities up to almost 70, dropping to 31 at the return of the campaign in April.

Although it is encouraging to confirm the power of the communications, this variation suggests that profound changes of attitude have not been achieved, or, at least, that the changes effected are not profound enough to be reliable. It could be argued that drivers are naturally reckless, and that only a repressive campaign based on fear of guilt and pain can contain them as long as it is always on; or, instead, it could be argued that we are not witnessing a positive campaign against the negative human nature, but a positive campaign against other messages that instigate people to adopt behaviours that are dangerous to society. This is the case in the film and television shows that glamorize violence and aggression, sensory overload, and present the unlimited freedom of the individual as a maximum goal.

Maybe an extended repetition of the TAC campaign would result in the long run in a change in cultural attitudes, as has been the case with smoking and hockey helmet use in Canada. Whereas 20 years ago, manliness in hockey was playing without a helmet, now the hockey helmet – much like the football helmet – is a symbol of manliness. A combination of public education, legislation, enforcement and good sense sometimes generate changes in attitudes, and in the cultural meaning of objects and behaviours. A change in behaviours is always welcome when one works in traffic safety or similar fields, but a change in attitude should be a long-term objective, so that the change in behaviours can be sustained and self-perpetuating. Attitudes are not to be defined as what people say they believe in, but as the beliefs that actually condition the actions of people.

The TAC campaign is extremely sucessful. There is much to learn from it. But we still have to look further into the problem of generating attitude changes, and into how to steer them towards the reduction of the many problems that confront humanity today.

See Figures 4.1 to 4.8 for a series of images from the TAC campaign, including examples from films produced after the initial set (names in brackets in the figure captions indicate copywriter, art director and director respectively).

Figure 4.1 *Country People Die* (Greg Harper, Rob Dow and Jamie Doolan)

Figure 4.2 *Tracey P-Plate* (Greg Harper, Stuart Byfield and Peter Schmidt)

4.3 BRITAIN'S HEALTH AND SOCIAL SECURITY FORMS

The design of forms has attracted a good many efforts since the early 1970s. From government and business officers to designers and plain-language advocates, many people have tried to improve the performance of form-users in their task of understanding and filling in forms. Graphic designers have frequently fallen into the danger of believing that a visually well-organized form will solve the problem. Plain-language advocates have run into a similar error, believing that a form written in plain language will make a substantial

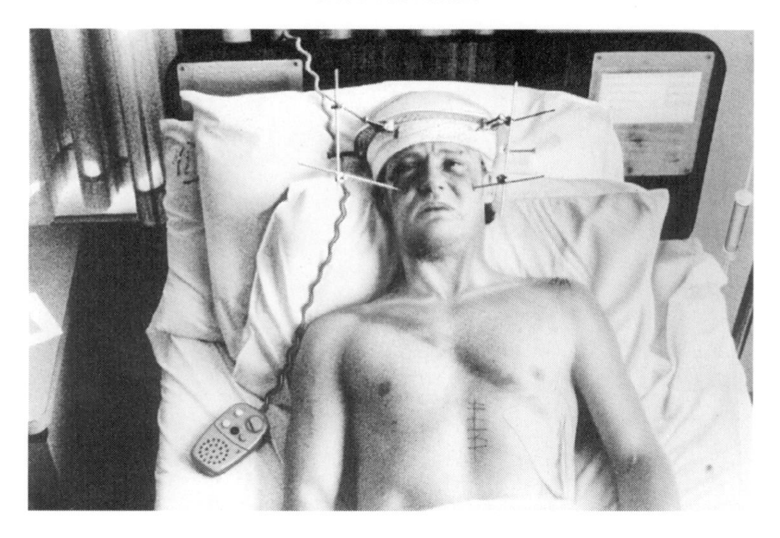

Figure 4.3 *Joey* (Greg Harper, Stuart Byfield and Rey Carlson)

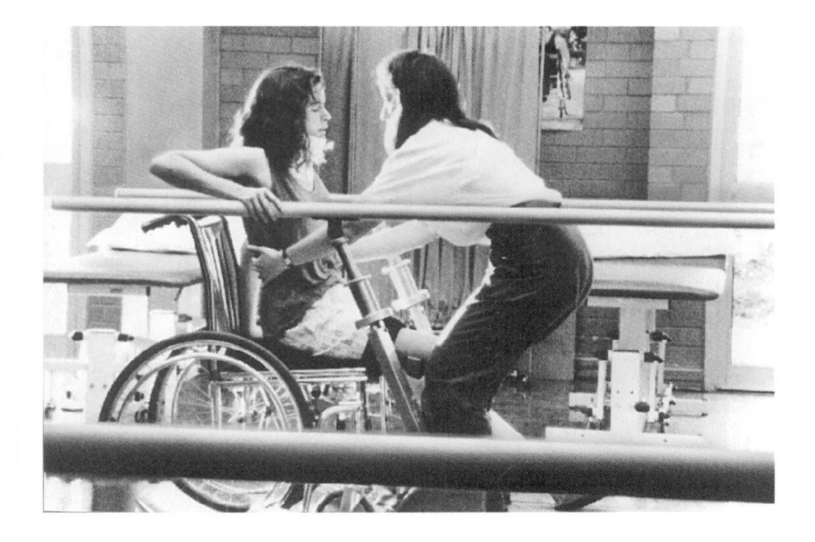

Figure 4.4 *Bones* (Greg Harper, Stuart Byfield and Rey Carlson)

difference; however, things are more complex than what any single-minded approach can conceive. A visually well-organized, well-written, good-looking and even well-liked form might not be good enough. Looks and the comprehension of individual bits of language do not necessarily predict the behaviour of the form-filler. In form design, 'good to fill' does not necessarily mean 'good to process'. The problems are many, and the approach taken these days tends to favour iterative methods, consisting of a long alternation of design and testing until a desired performance level is achieved.

Although traffic safety sounds like a far more socially relevant subject, the social and

Figure 4.5 *Country Kids* (Greg Harper, Stuart Byfield and Rey Carlson)

Figure 4.6 *Tracey P-Plate* (Greg Harper, Stuart Byfield and Peter Schmidt)

ideological implications of form design are many. Forms are the instruments through which many laws and civil rights are enacted. They are often the way in which government and people communicate. Joyce Epstein, of the British Research Institute for Consumer Affairs, wrote that 'Some one million old people legally entitled to public cash benefits are not receiving them' (Epstein, 1981, 215), beginning an article in which she reports on the problems related to communicating with the elderly, particularly in connection with the design of forms and in the context set by the opening sentence that centres on the rights of people to have access to what the law establishes.

One of the first consequences of the improvement of the forms used by the Department

Figure 4.7 *Bush Telegraph* (Greg Harper, Rob Dow and Mat Humphrey)

Figure 4.8 *Nightshift* (Greg Harper, Rob Dow and Mat Humphrey)

of Health and Social Security (DHSS) in Britain was an increase in the number applying for benefits. The difficulty of the previous forms had taken rights away from citizens, disrupting the two-way communication that they were supposed to facilitate. The redesign of the forms, begun in 1978 by the Salford Form Market, took some four years to complete, and involved a number of form developers and evaluators, including the Research Institute for Consumer Affairs of London and the Institute of Educational Technology at the Open University. Human factors specialists, plain-language professionals, psychologists and information designers worked to produce a series of new items.

Details of this process have been discussed at length in *Information Design Journal* (Cutts and Maher, 1981; Firth, 1981; Fisher and Sless, 1990; Lewis, 1986; Waller, 1981, 1985; and all the *IDJ* vols 2/3 and 4, 1981), with particular concentration on the analysis of the many factors that contribute to the functioning of a form. Initially, these factors can be grouped into two areas: a private person and an agency (government or business). However, although at the form-filler's end we can normally find just one person, at the form issuing and processing end we will normally find several parties, having different tasks to perform on the basis of the form, and diverse interests in it.

The problems of the form-filler are cognitive and emotional. Dealing with the restrictions of a rigid instrument can be difficult for a form-user who does not have a context for the questions confronted. In many cases, when forms have to do with the acquisition of benefits, the user is under stress. The combination of emotional stress and cognitive difficulties can easily prevent a person from obtaining what the law prescribes. The ethical importance of the redesign of the DHSS forms lies in this aspect. The success of the lengthy process was well worth the time and the effort, and stands as an example to follow everywhere. According to Robert Eagleson, the DHSS is estimated to have saved £13 million per annum on the processing of forms for an outlay of £350 000 in redesign costs (Eagleson, 1983).

At first glance it may seem to be an exaggeration but, according to David Lewis of the Document Design Unit of the DHSS of Britain, the DHSS uses a range of forms that vary in length from 2 to 80 questions, and in annual use from 73 000 to 4 133 000 occurrences. It becomes quite evident that errors in forms that are used 4.2 million times in a year can easily amount to a significant sum.

In a study of the cost of errors concerning 14 forms used by the DHSS, conducted by Coopers & Lybrand Associates for the DHSS, it was determined that the total cost of the remedial actions that had to be taken because of errors ranged between 1 000 and 1 137 000 pounds per form per year. In addition to the costs that affected the DHSS, the cost of errors to form-fillers was estimated to be between 400 000 and 1 104 000 per form per year. This cost estimate was based on the extra time used by form-fillers to remediate the errors at 50 per cent of the average hourly income, plus the delay in collection of the benefit. Errors that involved enquiries to employers also had an impact on employers' time, and resulted in a cost to them estimated as almost equal to the other two parties combined.

One can conclude that if the redesign of forms resulted in savings of £13 million per annum for the DHSS, the community also benefited from savings of several million pounds in form-fillers and employers' time.

4.4 AUSTRALIA'S CAPITA INSURANCE COMPANY FORMS

On a smaller scale of use, the Communication Research Institute of Australia conducted a redesign of a form for the Capita Insurance Company (Fisher and Sless, 1990). Using complex iterative design methods, they managed to reduce dramatically the errors as described in Table 4.1.

Table 4.1 Comparison of old form (A) and new form (B)

	(A)	(B)
Sample size	200	200
Percentage of forms containing errors	100%	15%
Total number of errors in sample	1560	44
Range of errors, form by form	1–27	1–4
Range of turnaround time to acceptance, in days	1–167	2–46

Based on an analysis of the cost of actions required to solve the problems created by errors made when using the previous form, the improvement in the performance can be quantified as in Table 4.2.

Table 4.2 Comparative cost of correcting errors between old form and new form

Old form (1987)	$551 464
New form (1988)	$15 441
Overall savings in 1988	$536 023

Although the size of the operation of Capita is significantly smaller than that of the British DHSS, the savings resulting from the project are clear and substantial. It can also be assumed that the form-fillers spent less time, suffered less of a nuisance, and understood better what they were doing.

The success of David Sless's work is based on the implementation of his convictions, already expressed several years ago:

> The designer has no alternative but to use more direct methods and get to know the user at first hand. This does not mean an abandonment of theory, which is always present when one collects information: the designer must choose theories which open up rather than close enquiry. Because this kind of work introduces a new and unsettling element into the designer's working routines, there is a temptation to hand it over to other 'professionals' but there are sound practical and economic reasons for not doing so. (Sless, 1985, 2)

4.5 BRITISH TELECOM: THE TELEPHONE BOOK PROJECT

British Telecom manufactures 24 million telephone books a year, consuming around 80 000 trees and 800 tonnes of glue. Colin Banks and John Miles, working with British Telecom, introduced a number of changes in the design of the book to reduce the use of materials and resources. This meant the design of a new typeface, with heavy strokes and clear endings and joints, which could withstand variable printing quality resulting from high-speed printing and could be sufficiently legible in a small size (finally used at 5.75 points). Also, type weight and width was reduced for addresses, since name-finding requires higher visibility than is required to read the text once the line has been found. This permits most entries to fit on one line. Particular attention was placed on the design of numerals and on stressing the distinctiveness of each letter, given the specific requirements of accurate name-finding tasks.

The page layout was changed from three to four columns, repeated last names were entered only once, resulting in both a saving of space and an increase of ease in the entry-finding task, because all that was left in a line was its unique features, i.e. initials or first names. Final savings in paper turned out to be in excess of 35 per cent, and the production cost for 24 million books per year was reduced by £6 million.

The positive impact is unquestionable: 28 000 trees are now saved per year thanks to the new design, resulting as well in savings at all levels of the transportation systems.

Although the positive impact of this project is to be praised, 52 000 trees are still used every year. The scale of the operation makes one confront the need to consider other avenues to solve the information accessibility problem addressed now by telephone books, and to look into some other form of solution. Although thinking about the need for 24 million 'smart phones' or home computers appears to be an environmental nightmare, the possibility that electronics offers to update the information without having to produce a new telephone book every year should be explored in its conceptual, ergonomic, ecological, economic and technological aspects. The objective would be to propose a design solution centred on the communicational need and not constrained by the medium currently used to address it.

This calls our attention again to the need to observe the design problem in the widest possible context, reviewing not only how a design solution can be improved within the existing paradigms but also analyzing the extent to which the existing paradigms actually represent an optimum response to the problem.

CHAPTER FIVE

Profiling the communication designer

5.1 INTRODUCTION

In this chapter, three colleagues reflect on the role of design, on the way it has evolved, on present needs and future possibilities, and on how design education could become an active agent for the development of a higher consciousness among communication designers about the possibilities of their action in society. These texts are not the result of a coordinating effort: their presence is intended to open the discourse, and to offer lines of thought that invite reflection and promote exchange. The attempt is not to produce the definitive profile of the designer, but to outline possibilities.

5.2 THE TRANSFORMATION OF DESIGN

Bernd Meurer

We experience the world in which we live through action, and through action we change it. Change implies creation or design (the German word *Entwurf* connotes both) even at the level of our everyday actions. In the specialized understanding of designers, however, the concept of change is usually limited to the design of products and their manufacture; it is understood to mean product innovation, a process that, in light of the distinction that our society makes between production and consumption, is considered completed when the finished product is made. This view of change, confined to the period in which the product is designed and produced, reduces the transformation of the lifeworld mostly to the creation of artifacts. Here, the lifeworld is viewed not so much as a process of civilization, but instead as a more or less (un)successful ensemble of physical and informational goods that are produced and consumed.

In this concept of design, the lifeworld is primarily conceived as a three-dimensional world. The temporal dimension is accorded a marginal role only. Even where the factor 'time' has always been taken into account, for example with regard to issues of perception and use, the temporal dimension is subordinated to the static concept of the 'finished' product, even in instances where the user interface functions as a dynamic information medium, or when air-conditioning, lights and other equipment constitute sensitive systems that respond to the presence and absence, as well as the movements, of people and things.

The lifeworld, however, can be experienced only as a temporal process that takes place spatially. The lifeworld is something that happens: it occurs through action and it is modelled on action. The lifeworld is more than matter that has solidified as form and in which time stands still. Its shape is defined through activity; action is its fulcrum.

By contrast, the way that architecture and design grasp the environment is fixated on things. Without a doubt, this fixation on objects in the field of design is concerned with processes that take place over time. But, in this case, action is usually reduced to the object-related function of usability, and not considered as a process of change and design in itself. In practice, designers seldom break through the constraints placed on them by this mode of thought.

If we construe design as being oriented toward action, and regard action as something more than passive use, but as active intervention and creative change, then design will no longer just focus on the object as a form. Rather, designers will be primarily concerned with how to develop and model processes: processes of interaction and change, in which objects nevertheless play an uncontested central role as a medium for action. Seen in this light, design relates to the entire physical and intellectual scope for interaction between people; between people, products and the lifeworld; and between products – in other words, between machines.

Today we must do more than merely refresh our awareness of these facts. Given the changes confronting our civilization, we must conceive them anew. Among the buzzwords for this process of change are globalization, economic structural change, increasing social inequality, ecological issues, changes in life and work habits, and the dissolution of traditional patterns of orientation, thought and action. In addition to the things that they actually stand for, these words suggest processes of change that neither proceed according to plan nor are completely random. Civilization changes beneath the surface of a wide variety of social, economic, scientific and creative designs through a kind of systemic chaos, where everything relates to everything else and is in flux. This shows that design change – in other words, change that is understood in design as calculated transformation – inextricably involves uncalculated self-transformation.

Action implies grasping, doubting, negotiating, deciding, altering and creating. Action is tied in with interests, and as such it is characterized by ambivalence, a propensity for conflict and ambiguity. Action is a communicative process. It takes place through motion: through intellectual motion, the motion of people, and through the motion and reshaping of knowledge, substances, things and data. Motion is a spatial process that takes place over time, although the temporal side to electronic data transfer eludes sensory perception. Given the differences in technologies and speeds, we distinguish between two kinds of motion: transport and communications – two concepts that were in no way separate until late in the nineteenth century.

The division of telecommunications from transport began as early as the age of smoke signals. However, it did not become firmly established until the electric telegraph was introduced in the nineteenth century. In those days, the term 'communications' still indicated any and all kinds of exchange; in other words, it included traffic and transportation systems too. At the beginning of the nineteenth century, transport and telecommunications were still one and the same thing. Disregarding techniques such as semaphore, for example, information was transported in the same way as goods and people. The interrelation of time and space still played a central role in passing on information.

In today's telecommunications, spatial distance is hardly of any importance at all – at least on a terrestrial scale. Thanks to telecommunications, people who communicate with each other but are in separate locations are temporarily brought together. We can move about via the network of virtual space without leaving our chairs. What we perceive as telereality and where we are when we perceive it are co-temporal. Space and time appear to part ways.

And yet within the sensory perception of telereality, space and time, as in reality itself, together create a space–time continuum. For example, the 'Terravision' project by Art+Com in Berlin, albeit still in its infancy, is a virtual reproduction of the planet and potentially everything that happens on it, a real-time copy of the actual world. Via such systems, the individual, as a 'user' with a 'space-mouse', can enter the virtual space–time replica of the real world, and thereby attend a concert in Moscow or a scientific discussion in Cambridge, or watch the construction work on Potsdamer Platz in Berlin. Just as easily – and this is the other side of the coin – the user can flick through these worlds independently of their spatio-temporal context.

The expansion of the electronic network serves to transform the shape of communicative action and the possibilities for it. Buzzwords such as telework, telebanking, tele-education, teleshopping and telemedicine are examples of the extent to which traditional concepts of space and patterns for action have dissolved in everyday life. In the near future, we will have access to 15,000 television channels. The quantity of data available is growing, as are the data flows. Unlike the printed word, it seems as if there were no technical limits to electronic storage and distribution of information. For quite some time now, the problem is no longer a lack of information, but rather a glut of it.

The moment when everyday life became 'opaque' has long since arrived. Without orientation systems to guide us, we are lost. And not only in traffic, in the subway, or at the airport, but also – and especially – when it comes to digitalized information processing. So-called 'personal information management systems' are being provided to handle the flood of information; however, developing these management systems and processing information involves more than just making information available and clear. It is not enough simply to handle the technical aspects of the microprocessing network, which is growing by leaps and bounds. Rather, we must also understand the impact of this phenomenon on our civilization. Today, everyone who is logged into the network can input or download any kind of information. A central issue here is whether or not people are capable of penetrating the information environment on their own initiative, dealing with information responsibly and competently.

By introducing microprocessors at all levels of our everyday lives, we have altered our understanding of use and the way in which we use things. Increasingly, 'use' is changing from being a 'hands-on activity' to teleintervention. The form of those products that we still handle directly is conditioned by the hand or the human body, in addition to other factors, such as their function, their technology, and how we perceive them. Teleintervention implies controlling processes via a medium instead of handling the objects directly. An increasing number of products can be equipped with features that respond to the user by implementing new technologies, from the fields of robotics, self-diagnosis and image recognition, for example. Not only are these sensitive products capable of responding to external control impulses: they are also increasingly able to control and transform themselves. Objects and space become sensitive, interactive entities that respond on their own to brightness, colour, temperature, moisture, movement and sound, and thereby interact both with us and with each other. The intrusion of robots into everyday life – similar to developments in the media – can potentially exacerbate cultural poverty. This, albeit unintentionally, illustrates Nicholas Negroponte's description of the networked household, in which music follows us from room to room, lights turn on and off by themselves, and appliances communicate with each other in a fully automated kitchen, drawing the car, the garage door and the alarm system into a conversation in which the refrigerator tells the car that the milk is gone, and the latest stock market quotes are etched by heat into the toast at breakfast. Machines will learn to recognize individuals by their voices, gestures and facial features, and literally learn to read their lips. Microchips will also find their way into clothing. In addition to clothing us, these 'smart' garments will also function as IDs and appointment calendars. According to a study carried out by Creapole ESDI in Paris, they will in effect be 'wearable computers' that receive, send and process information; furthermore, they will be able to change their colour, texture and breathability on their own. Clothing will become part of the network as a body shell. The social, cultural and ecological problems unleashed by these developments reveal how necessary it is to rethink our concept of use and ease. The easier something is to use, the less we think about how we are using it, and which 'side effects' such use implies. The more perfectly a product has been designed, the less we are tempted to consider any problems posed by it or its use.

As robots begin to belong to our everyday lives, design will be forced to confront the issue of processes, and how these processes take place. Products will soon have a self-transforming character, although their scope for doing so is still limited. They are acquiring the features of the machines that manufacture them. Thus, processes that were previously limited to production are finding their way into the domain of consumption. The production process is beginning to extend beyond what is normally associated with manufacturing, with no clear limits in sight. The old adage that production finds its completion in consumption, takes on a whole new meaning here. Use and transformation interweave and affect one another in new ways.

Products are viewed as processes in ecology as well, but for different reasons. This applies both to the ecological cycle of a product, from its manufacture to its disposal, and to the product's function as a medium for action and economic commodity. The Wuppertal Institute for Climatology, Environment and Energy draws up an ecological 'balance sheet' for objects and processes by analyzing all the relevant equipment, raw materials and products, and the converting, packaging and transportation processes. Their investigation addresses all aspects of an object or a process as it emerges, during its existence, and after its disposal. This extends to tapping raw materials and energy, production, use and maintenance, as well as all recycling, disposal and decay processes. The materials flow that results from these processes is considered to be the decisive factor in determining to what extent a product burdens the environment. 'Materials flow' here means the extraction, transport and transformation of all substances that go into the manufacture and use of a product, as well as all subsequent processes.

The goal is to reduce materials flow and thereby lessen the damage to the environment. This can be achieved, on the one hand, by increasing material intensity in the product life cycle, in other words by reducing the quantity of materials used while maintaining or enhancing performance. On the other hand, the hope is that by applying service-oriented economic principles to our supply economy, with its emphasis on boosting unit sales, products will be used more intensively, thereby braking the upward trend in product quantities without sacrifices at the consumer end. Here, change seems to be occurring. Not for ecological reasons, but mainly for economic reasons, service-oriented approaches are being developed in certain areas of the supply economy. In this context, a service orientation means uncoupling the use of certain products from their ownership. Rented apartments and rental cars are long-standing examples of this principle. In the transportation industry, for instance, 'individual mobility' is being considered independently from private car ownership. This is conceived as a service that must be readily available as well as highly attractive for potential consumers. Car sharing and car-pooling are mere beginnings of this new conception of individual mobility as a service where products could be used more intensively. Ownership and use do not necessarily have to be one and the same. This is especially true of software. Products dissolve into services; processes come to the fore. Telecommunications form the basis for service-oriented forms of the economy, with networking as the key factor.

In practice, interest in networked computer technology within the field of design primarily focuses on image generation; however, the radical changes that will prove decisive for the course taken by civilization are occurring on a completely different level, one that is not visual and appears far less spectacular on the outside. As the networking of computer, telephone and television technology continues to unfold with little fuss but great capital outlay, it is spawning a global arena for action that defines and incorporates individuals' actions on an economic and social level – and above all in the world of work – in a completely new way. Work, which it was possible to organize rationally in the mechanical era only in large working communities, is increasingly becoming part of the virtual network. When the work world is finally networked on a global scale, it will become a comprehensively integrated institution, not only for those on the network, but also for everyone else, provided that they do not fall through the meshes. In this new, synthetic work world, every form of work that can be executed and transmitted via computer technology can also potentially be called up and carried out by anybody anywhere in the world, 'just in time'. Not only jobs, but also study places will no longer be tied down to central, collective workplaces.

The networking of work changes the social and spatial organization of work, and thus the social and spatial organization of the lifeworld. This presents completely new challenges for design. In organizational terms, work is being networked on various levels, one of which is called telework. In the European Union, there are at present around 200 000 telework-stations. In the USA this figure is already eight million. In Germany, according to an estimate by the National Association of German Industry (BDI), in just a few years there will be more than three million telework-stations. The traditional forms of work that are tied to common places, such as factories or offices, lose their significance. This affects human social behaviour, the spatial organization of cities, and the typology of buildings too. The traditional spatial division of cities into separate residential, work, consumer, educational and resort areas has its roots in the principles of the industrial division of labour, and developed historically in that context. The tradition that has shaped urban space, in which people leave the residential areas in the morning to gather far away in commercial districts to work in a joint setting, and then in the evenings return to their homes, is barely 150 years old. It is a product of the principles of the industrial labour society, through which the cold wind of structural change has been blowing for a long time. In historical terms, the city divisions based on these principles will certainly not be the 'ultimate' shape taken by urban organization.

Work is becoming an endeavour that can be carried out independently of location. It is becoming literally mobile, which in turn corresponds to the increasing demand for mobility in the job market. The local job market is becoming global. Not only are we able to transfer manufacturing to countries with lower wage levels; in addition, administrative, scientific, design and cultural work is being made available on the information superhighway. Already, according to a statement made by the Institute for Economics and Society, highly qualified workers in low-wage countries such as India are writing computer programs for multinational companies. Recently, there was a statement in the press that Mercedes-Benz is building a research and development centre in India. Networked work means more than exchanging information at the speed of light, information that could just as easily be sent in printed form. The crucial difference is that in networked work, computer-generated data and constructs are no longer processed separately in individual computers and then made accessible to everyone else; instead, it is possible to develop them in a linked network of computers that supplement and enhance one another.

Take Otis, the American elevator company. This company owns production plants at 24 locations across the globe, and conducts research and development at 15 facilities worldwide. At present, the work technology used by these development groups links them as part of a global computer network; they are then supposed to work together as a so-called 'cosmopolitan development crew' – cooperation that takes place seamlessly despite time-zones. The research day is extended to a 24-hour workday. Shift work mutates into time zone work. Within these cosmopolitan work groups, the structures of real, local work communities are coupled with supra-local and virtual structures, and are thus less binding in social terms.

In networked work, the individual acts in real time both in network cyberspace and in real space, to which he or she remains physically leashed, with everything that this implies. People speak mystically of bodies that vanish in cyberspace; yet, at the same time, the human body has never been given as much attention as a real entity as it has in our times. Even the fitness industry is booming.

The ever-expanding information networking also has an impact on the status of transportation. As machines aimed at saving time, means of transportation cannot keep pace with data transfer technologies; however, as machines for transporting people and goods, they are in no way losing importance. The Frauenhofer Institute for System Technology and Innovation Research maintains that the advent of telework will reduce professional and business traffic from 34 per cent today to 28 per cent. However, there is no reason to suppose that in this context traffic growth will diminish. After all, it is modern telecommunications in the first place that allows people to go anywhere at any time. In addition, the new technologies will also make the workplace itself mobile. The information highway will in no way ease the burden on the real highway.

The meaning of space, time, motion and speed is changing. Two apparently paradoxical observations summarize the present situation of the urban space. On the one hand, people are speculating that better and faster communications technology will make urban communal life superfluous, at least in communications. On the other hand, urban conglomerations are expanding at an alarming rate. When Paul Virilio says that the city is no longer necessary because people no longer require spatial proximity to each other to establish linguistic and visual contact, this merely specifies that in purely technical terms it would be possible to communicate without ever leaving our houses. However, we can hardly expect that communication will be reduced to this level through the development of telecommunications technologies.

The social structure is what is changing, together with the spatial organization of urban life and business. The virtual, global economic network – the 'global city', as Saskia Sassen calls it – is being superimposed on real urban space. Unlike in earlier times, a city's economic importance will no longer be defined solely through its function as a regional hub for the exchange of goods and information; rather, the function that it assumes in the international economic and financial network is becoming the crucial factor. Thus, a gap is emerging between a city's economic function in the international economic network, on the one hand, and the actions of the city dwellers in their local context, on the other, severing the connection between urban communal dwelling and economic action, which is what creates identity. This development has a correlate in the increasing social inequality within cities. In Germany, which certainly cannot be considered a poor country given its GNP, the bottom 50 per cent of households possess only 2.5 per cent of all monetary assets, whereas the top 10 per cent possess more than 50 per cent. Alain Touraine refers to this phenomenon as the increasing presence of the Third World within the First World.

Along with the upheavals in the work world, the structures of social space are also changing. As work is increasingly a matter of microprocessors, the economic foundations for the social system that developed over the last 100 years, based on a form of gainful employment in which the employee is directly dependent on his or her employer, are being eroded. The social security systems of the old industrial society are unsustainable. New systems need to be invented. At the same time, communal forms of work in many fields are being replaced by workstations that are networked and not in the same physical location. As the form of collective work spawned by the industrial society vanishes, this society also loses its potential to engender communities. Customary orientation schemes, principles of behaviour and social safety nets are no longer effective. Something else arises to take their place. The question is still open, however, as to which new forms of community-oriented action could be developed on what basis, and how. On the one hand, this vacuum provides historical scope for developing new options; on the other, it provides the pretext for clinging to purportedly proven concepts with fundamentalist zeal.

With the mobilization of work (buzzword: telework) and the globalization of the economy (buzzword: separating economic processes from the workplace), the basis for the social organization of our lifeworld is undergoing a transformation. Ways of life are changing, and so are the demands made of design. It would be more than daring to try to predict what will take their place. Instead, the crucial issue is to broaden the concept of design to accommodate new challenges, and to provide arguments that will free it up as a social, political, economic and cultural scope of action.

Today design must be construed as sustainable development. Sustainable development means more than environmental design with a focus on ecological aspects. According to the words of the Rio Earth Summit, 'sustainable development' claims to link 'ecological sustainability' with 'social equity' and 'economic development'. Design that intends to rise to meet this challenge is all-encompassing, tying in ecological and cultural aspects as well as spatial and temporal ones. The interdisciplinary approach, however, as it has been practised to date, has been incapable of effectively combating the problems generated by dividing work into separate spheres of expertise. Instead of just adding disciplines to one another, there is a need to develop work principles that prove more effective to integrate them with one another. In doing so, specialist know-how will be irreplaceable. Design that is oriented toward sustainable development is a complex of activities that transcends typical images of professions. It can be neither conceived nor taught as a separate field. It must be developed intersecting existing disciplines and individual areas, within the scope of projects and flexible groupings constituted and adapted specifically for the problem at hand. To achieve this, a new kind of institution must be established: facilities that are understood as an open, interactive field for design, science and technology. Design must be understood as a networked activity, based on the model of specializing according to the given context. This does not mean that the various disciplines are to be done away with. Instead, the nature of specialized work must be changed by orienting it to specific problems rather than to traditional disciplinary areas. Of course, this presupposes disciplinary competence.

The question needs to be raised once again: what constitutes design, and which criteria will be used for identifying and questioning design issues? Design, science and technology have to see themselves and each other as objects of creative reflection and intervention. This raises doubts on whether or not design, science and technology – areas of activity that, as history demonstrates, are capable of changing the lifeworld from the bottom up – are capable of transforming themselves. Thus, we must challenge the social organization of design, science and technology, a concept that is at odds with the established distinctions between spheres and disciplines, as well as their resources – particularly at universities.

The concept of superimposing various disciplines to address the problem or project in question could spawn a new hybrid category of design activity, which will emancipate itself from traditional disciplinary concepts. It would be equally oriented toward research and practice, albeit in a novel way. Design that acts responsibly concerning the future, that is critical and analytical, that asks questions and develops alternatives, that discovers causes and contexts, and that develops new, comprehensive ways of identifying problems and forms of design action, demands that both scientific concepts and shaping concepts for change be developed so that each complements the other.

Design is impaired above all because many people, both designers and non-designers, see design as the creation of unquestionable answers. In this context, sociologist Ulrich Beck speaks of 'the culture of certainty' and its fateful role. Undoubtedly, the search for absolutes is an age-old phenomenon. But certainty excludes doubt, and thereby turns into dogma. Doubt goes hand in hand with uncertainty. And doubting what exists, what one thinks and what one does, is the most important aspect of creative action. Doubt is a prerequisite for creativity. Design should be effective, but in its effectiveness it must also see itself as a self-created problem. It must acknowledge the fundamental doubtfulness of its own action, and create a public awareness of this.

Design must be liberated from the one-dimensional mode of thought that focuses on solving tasks, and instead it must be seen as the constant creation of new tasks.

5.3 DESCHOOLING AND LEARNING IN DESIGN EDUCATION

Jan van Toorn

Design education should deschool but at the same time enable one to learn by experience. Like all other education, it should be a pedagogy that contributes to the independent development of the individual and to a free practice. And it should generate a serious engagement in establishing a liberating force in the formation of public opinion.

I have chosen this formulation because the present role of design is the complete opposite. The idea of utilitarian modernism that the designer serves the common interest by creating a consensus between producer and user is increasingly being superseded. If we did not know it already, we have had to learn that intellectual mediation goes beyond the organization of use: the introduction of tools and methods has an autonomous effect on social reality. The capitalist economy has in this way created unprecedented opportunities for design, and in turn design has become an indispensable factor in the research, development and distribution of the products of the (culture) industry. As a result, and with the help of this central role of design, our everyday experience has become more and more aestheticized by the symbolic violence of neo-liberal production and circulation. Goods and information have become permeated by convenient stereotypes that block socio-political conceptualization. Technology and aesthetics create fascinating entertainment, which serves to reassure the consumer – who 'elects freely' – in the confrontation with the problems of the natural and social environment. Through this symbolic mediation, individual thought and the public climate are at a crippling impasse.

A liberating pedagogy should recover the forms of making and thinking, which have been instrumentalized by trade and services, so that they become useful again in the search for conditions that allow a more truly human existence. This requires a consciousness that differs from the dominant paradigm of most practical intellectuals such as designers. As far as I am concerned, this is only possible from a social–libertarian perspective which does not view the social and symbolic order as immutable essentials. They are temporary and historically determined constructions which must be revised continually in the light of current circumstances.

Schools provide an excellent space for such reflexive and operational research. It is precisely on the periphery of the worldwide media spectacle that there is room to develop an alternative practice. Students should learn to face the challenge of change because the school offers them the means to perceive, reflect, criticize and transform.

Design education, if it is to contribute to a more radical democratic formation of public opinion, must go further. It also involves accepting the paradoxical position of the mediating intellectual whose actions are embedded in an irreconcilable conflict of interests. It must contrast the established image of the professional who operates in a complete and coherent system of norms and rules with the communicative reality. The history and discourse of the discipline in its dealings with production must be related to the social factors which cannot be considered separately. It is only by relating professional practice to its conditions and consequences that the designer is forced to anticipate a use not laid down by tradition. This alone creates the opportunity to act in a partisan, non-authoritarian, specialist fashion.

For the school, this means a programmatic attitude that consciously places the action of the 'pupil' into a dynamic relationship: in other words, within the complicated contradictions of private and public interests. To formulate it precisely: the student must learn to make choices and to act without attempting to avoid the tensions between individual freedom, disciplinary discourse and public interest. That is not only a plea for a mentality that allows the student to survive both creatively and economically but also, and above all, a 'survival kit' that makes it possible for him or her to develop heretical strategies within the dynamics of this trialectical relationship – strategies that enable the addressees to become more independent in their dealings with the message.

This does not mean that the school programme should not focus on the growth of individual expression. On the contrary, the mutual presence of students and teachers must lead to a qualitative improvement in their visual literacy. The only way to attain that is by training them in the achievements of the professional realm. Without the experience and tools developed in this historical space, we are, at the least, seriously handicapped. Dependent as we are on the linguistic and methodological references of the discipline, the school is an important instrument for deepening, broadening and actualizing these references by personal action and reflection, which is quite separate from our desire to reject, transform or maintain the specialist order.

The social space that has evolved in time to accommodate the development of the disciplinary field would have been impossible without the help of other groups in society that created the material and spiritual conditions. It is in reference to this social dimension – to the disciplinary space itself, or perhaps, even more accurately, to the relationship with the other mediating groups – that the designer's (the intellectual's) nature is exposed. There is, on the one hand, the claim upon autonomous symbolic production and independent public service, and on the other, the dependency upon common symbolic references and social setting without which those services could not exist. This is a complicating factor, and conflicts with the often 'subversive' pretensions of individual designers and the humanist tradition of the profession. Designers who do not recognize the insurmountable conflict of interests in which they work, lose entirely any possibility of liberating the disciplinary practice from its restrictive institutional character. By focusing during schooling on the conceptual history of design in relation to the economic and technological development of production and distribution, the school contributes to the student's strategic insight. That understanding of the way that design responds to changing relationships immediately raises questions about the meaning and structure of language and the communicative consequences thereof in the social context. And it familiarizes students with the artificial nature of symbolic production – an experience that challenges each of them to manipulate the forms of expression of the disciplinary sanctuaries according to their own views.

The impulses that feed our individual freedom and safeguard against the gloss of the instrumentalized rhetoric of the disciplinary realm are in essence the results of our experiences in the real world. This notion enlarges the social dimension of the school's programme by considering the tense relationship between the prevalent image of the world and the factual social and natural situation we live in. The intellectual paradox of the disciplinary space becomes even deeper and more uncomfortable to deal with here. Whether we want it or not, this dimension is saturated by our engagement in the orthodoxy of the dominant culture. Chained to that bulwark, the quality of our individual independence is determined by how much we can liberate ourselves from it. This is all the more oppressive because the images of reality are increasingly the result of corporate mediation, including that of design. Within unbridled capitalism, the culture industry has divided the public sphere almost entirely between interlinked private combines. A far-reaching economic core extends to every aspect of culture and has made deep inroads into daily life. This expresses itself in the production of superficial generalizations of aesthetic modernism, in which objects and information have lost both meaning and origin. The unavoidable result is that much of what design produces is self-referential and serves to reassure, rather than to expose, a more truthful experience of the world.

Design education with pretensions for liberating the pictorial and discursive formations of the public sphere should do so to improve the conditions for social change. It should offer orientations and arguments stimulating our human ability to deal creatively with reality to open up and to transform the symbolic conditions of the existing culture, which means that the students will find other spaces to act, and consequently other themes and more adequate forms of manipulating the vocabularies.

It seems to me that design can no longer claim a representational and interpretive monopoly of our experience. The issue now is, to quote Pierre Bourdieu, 'the transformation of [specialist] field structures' by investing in the analysis of our collective circumstances. Now, more than ever, that is by no means a simple task, because there is an enormous lack of cultural and political orientation – of obstinate thinking and action on the democratic quality of the public climate, which is in a deep crisis.

In such a transitional period, it takes courage to combine the factual and symbolic uncertainty of the social space with an investment in individual professional action. There is a real danger that greater insight into the fundamental schizophrenia of the intellectual will lead to cynicism and inactivity among students. Deschooling is therefore above all to learn by making, by operational action. Deschooling, which examines the nature of the problems, should enable students to tackle them, after all, with more success than achieved previously. It is by action alone that the student will be able to acquire an operational relationship to his or her social options.

Critical explorations that go beyond the articulated formal consciousness of the disciplines remain dependent on the way that we deal with our paradoxical nature as intellectuals. This applies not only to individuals but also to disciplinary domains and professional training. For me, the challenge and the pleasure of exploring concrete possibilities with students for a real, independent and more liberating design practice are the most important reasons for my association with the school.

5.4 DESIGN PRACTICE AND EDUCATION: MOVING BEYOND THE BAUHAUS MODEL

Dietmar Winkler

The Bauhaus was perhaps elevated to symbolic status, by an enthusiastic and emerging American design practice and its supporting design schools, as one of the first Western institutions representing democratic ideals and modernism. Design practitioners, becoming more intertwined with the industrial and corporate world, saw in Bauhaus a good grounding for their professional identity and for breaking away from the domineering self-expressive arts. The Bauhaus provided them and design educators with a rational form and colour language, which was a welcome set of tools when efficiently linked to practical concerns in the argumentation for the effectiveness of visual concepts and solutions in front of clients and critics. They readily took advantage of the opportunity.

Although the *beaux-arts* had lent much to early design and art education, one fact is clear: the Bauhaus educational rhetoric and connected value system were adopted uniformly into American design education. It was considered not just as one way to investigate visual phenomena but also as a superior way. It is because of this elitist perception that the Bauhaus's hold on design curricula has become phenomenal. Its contributions are lauded in most design textbooks; its pedagogy is considered infallible.

But for graphic designers, the Bauhaus's narrow investigation of principles of colour and form creates a distinct dilemma. Design, when it is at its best, is inextricably linked to the processes of communication. It functions best, in facilitating communication, when designers, as arbitrators, reconcile the various differing needs of client and audience. To design well means to understand the complex interactions of human contexts in the communication environment. It requires awareness not just of visual perception and visual discrimination but also of the total ecology of valuation, value discrimination, identity, territoriality, status, and anything that influences and alters humans' social and personal behaviour. In their practical preparation, designers face curricula that are rarely responsible for investigations beyond formal aspects. Most courses are seriously deficient in addressing social or behavioural issues reflected in complex communication environments. Although *beaux-arts* schools fostered intuitive and self-expressive responses to the visual environment and the Bauhaus made logical analysis the centre of visual investigations, neither of the pedagogies, however, considered the human interface to visual images. Neither of them considered design as a primary support function to communication, responsible to social and behavioural contexts.

Looking at American design education, most schools' curricula are rooted in the Bauhaus and its model of design with emphasis on formal aspects: hand and technical skills. Visualizing skills are still considered as most essential. They are the outside measure and manifestation of the graphic designer's physical control over materials, methods and reproduction technology. They show the designer's visual perceptual prowess. This concentrated focus on visualization skills overshadows the students' intellectual and cognitive development. If design practice is to emerge as a recognized profession, it must overcome the barriers of the intellectually limiting Bauhaus model. Like other professions, it must begin to develop a strong information and knowledge base founded on contributions to a body of original research. The speed with which the design profession will surface is unquestionably linked to the agility and ability with which designers and design educators build a reliable and verifiable information reservoir. The design practice must begin to recognize the need and to support the idea of research and testing. Design education must expand its undergraduate curricula to include cognitive studies and the introduction to research methodology, and human, social and environmental factors. Graduate education must be reorganized to take original research as a primary requirement, moving away from the continued emphasis on vocational skills, and to provide the field with useful information.

In the American consciousness, the Bauhaus took on an aura of moral and ethical inspiration. Its founders were perceived to have left Germany in defiance of Fascism, and it provided practitioners with the tools and rhetoric of formal design aspects. Today, the rightness of the Bauhaus's principles has given way to doubt. One must be critical of attitudes that, instead of being truly international, impose foreign concepts on local culture, design education and practice. These artificially injected values interfere with and destroy the colloquial and vernacular expressions of an existing culture. Many of the ideas that evolved in the Bauhaus are an integral part of the American design establishment, which has grown from a handful of designers in the 1930s into a mega-institution. There, the ideas stand unchallenged as part of its fibre and politics, but the reason for their existence has become vague. It is likely that the original intent of Bauhaus teachings was honorable, but turbulent times carried it down a path that it might not otherwise have travelled. Its vision was deceptively new and the narrowness of its educational and ethical goals hidden. The emerging challenge to the Americans who are true believers in the Bauhaus is to make room for those interested in building the intellectual foundation for a local design profession. Design will not be able to provide cultural leadership unless it is recognized and respected as a knowledgeable and equal participant in the dialogue with other disciplines to evolve better human, cultural and technological futures. The Bauhaus is partly responsible for the myths that it created, but the blame for glossing over its defects must be shared by those who unquestioningly appropriated its dogma, or who bent the truth to their own political needs, or who benefited from association with its purported ideals.

To blame the Bauhaus as the primary reason for the lack of intellectual focus in American design education is not totally fair. Up to the 1960s, and even beyond, the field has been well served by the Bauhaus model. Designers enjoyed their separation from advertising, the change of nomenclature from commercial art to design, and the explorations of the Bauhaus form ideology helped their emerging prestige and confidence. The field was sustained by the limited focus. In today's period of dynamic change affecting all aspects of life, however, the field faces design problems of greater complexities and ramifications, and it should venture into intellectual and context-based design.

The Bauhaus was in essence a typical German *Fachschule*, a school preparing students for vocational practice – no more, no less. The students' education was seriously lacking in intellectual stimuli: no theory was taught, no economic or political history, no natural or social sciences, no music, no formal art history, and no literature. This educational philosophy denied academic pursuit. Exploration of intellectual issues was encouraged only if it affected the craft, its process or its materials. This lack of intellectualism should not be surprising. Gropius and the directors who followed him, Hannes Meyer in 1928 and Mies van der Rohe in 1930, had been steeped in the trade-school tradition, which saw non-applied research and intellectual pursuit as the dilettante activity of the rich and aristocratic. Unfortunately, however, the Bauhaus faculty did not recognize the restrictions of their own self-imposed straitjacket.

This was aggravated by a gradual trend toward anti-intellectualism that occurred in Germany over the course of almost 40 years. Beginning around the turn of the century, it emerged among the Junkers – the landowning nobility and the old-guard military. Anti-intellectualism, consisting of a twisted defence of German values, surged again during Hitler's dictatorship in response to the rapid growth of industrialization and the consequent shift of political power from the country to the city and town. The landowners saw themselves in sudden competition with industry for cheap labour.

The Bauhaus naively participated in the perpetuation of the German class structure by continuing the usual distinctions between the working class, the educated, the politically powerful and the affluent, and by presuming to prescribe a lifestyle for the working class: high-rise housing and mass-produced furniture, wallpapers, textiles and kitchen utensils. Although the school wanted to be perceived as having a democratic view of society, in fact it imposed its ideology without consultation with or concern for those who had to live with its experiments. It did not question the impact of its design on the users, whose agreement was simply taken for granted.

Long before the Ulm school organized one of the first German-produced design curricula, the Bauhaus recognized in the 1920s the significance of the increasing industrialization of German manufacturing and made a distinction between mass production and the craft of producing one-of-a-kind objects. Later, the main studio activity was directed towards mass production. This interest in industrialization was genuine, the time was right, and this central focus distinguished the Bauhaus from and gave it an advantage over the Academy and the other craft schools. Josef Albers, who taught a course in the properties of materials used for fabrication, said that, at the Bauhaus, they never used the word 'education' once. They spoke about influencing industry.

The Bauhaus's approach to functional design, however, which dealt only with manufacturing technology and never even touched on the problems of ecology or the personal complexity of German life, led the Bauhaus to an untenable position – the assumption that all Germans adhered to the same lifestyles and values.

Although the Bauhaus commitment to addressing the problems of visual design created by technological advances was invaluable, particularly during the period of post-First World War industrialization vital to the reconstruction of Germany, its imposition of its own creative and social ideology was ultimately self-defeating, contributing to abundant critical response and waning interest in the Bauhaus. Credos about good design did not inspire the public. The worker's dream was a big art nouveau mansion like the industrialist's, not a flat-roofed, white box. For him there was no cultural metaphor for the new architectural language other than the army barracks, which every German male knew well, or the Calvinist churches, or government buildings lean in embellishment or ornamentation.

Looking at the traditional German architecture of any region, one recognizes a wealth of visual expression: from the Baroque in Catholic regions to the austere in Protestant areas, there are many styles, each culturally supported by its constituency, each reflecting the regional identity. The Bauhaus ignored these common cultural traditions.

In the absence of populist cultural perspective, the Bauhaus developed an educational model outside the communication experience of the majority. Certainly, it could not be deciphered by the public for whom it was intended. The Bauhaus asserted itself in the value of invention, or did little to restrain eager followers from allotting it more credit than history would dare. The narrow Bauhaus point of view imported to the USA competed uncomfortably with the multi-faceted, anti-socialist local spirit. Europeans were also confused by the constantly shifting and elusive democratic process as it affected issues of individual and collective rights. The American dream of success and leisure reflected in media presentations of Hollywood-style houses and heart-shaped swimming pools appealed on a broad scale regardless of cultural, ethnic or economic background, or geographical location.

The New Bauhaus in Chicago, expanding its roots, was primarily interested in reaching a zenith of the same renown as experienced before in Germany, seeking appreciation from a cultural elite, a minority of the educated and sophisticated, and the corporate sector. There were few drastic changes in the curriculum. The studio skills again outranked the cognitive skills. Nagy, and later Kepes at MIT, gave intellectual ambience to their institutions through publishing and editing of publications. In reality, there was a lot less going on in the everyday classroom. Image and object making was considered by the institutions as part of the arts and was therefore far removed from other disciplines. Only under rare and fortuitous circumstances did syntheses between different departments occur.

By buying into the Bauhaus model, the American design institutions also bought into its limitations. The Bauhaus faculty was absorbed into Ivy League schools at an opportune time. The greater American national need for sophisticated technology, before 1945, to support the Allied war efforts began the integration process of applied research into the American Academy. The blur of distinctions between liberal and vocational education was further intensified after the end of the war. This historical twist brought the vocational guild system of the Bauhaus in touch with academic attitudes, forming a new tradition. Today, even the Ivy League schools support vocational graduate as well as undergraduate education in which the academic portion is a mere fraction of total graduation requirements.

The integration and acceptance of the Bauhaus's vocational design education model by colleges and universities was furthered by Gropius's appointment to Harvard University. Harvard's programmes were set up not for the vocationally trained, but intended as a continuation of education for those from other Ivy League schools, such as Brown, Yale and Princeton. Graduate students shared the common experience of a well-rounded humanistic education, and at the same time membership in the established, secure and educated upper class. Their common humanistic experience made it proper for the university to support a professional focus for graduate education. The limited academic requirements and proportionally larger studio segment of their programmes were justified because each student was aware of the value of academic pursuit. In a long-standing intellectual tradition, Harvard graduate students understood that to hold future leadership positions they had to continue their humanistic studies. Completing doctorates in other disciplines was always part of their considerations.

There was a flourishing of new design programmes, especially at state-funded schools, after 1945. Schools responded to the cultural and economic shifts from agriculture to industry, the growing social changes initiated by flight from rural areas to the cities, and the growing human aggregate's needs for support services. The industrial and commercial demands for trained personnel, and the desire to strengthen the supporting service system, unfortunately helped to speed up the misinterpretation of the role of academic studies in graphic design. Copying Harvard's model of graduate education, but eliminating the strong humanistic undergraduate base by substituting vocational components and adopting the Bauhaus's non-intellectual focus, has shortchanged both design education and practice.

In the competitive, institutional accreditation process for programme credentialing, self-serving schools have successfully persuaded American accrediting bodies that undergraduate design education requires a two-thirds studio emphasis and, even more destructively to the professional development, that it is appropriate for graduate design education to continue emphasizing hand and visualization skills. It endorses the graduate programmes' small percentage of academic courses and reinforces design's institutional disrespect for original research.

With this narrowness, however, design practice cannot support its professional ambitions; it can barely support itself as a vocational craft. If the model of design as an aesthetic beautification process stays unchallenged, then anti-intellectualism is institutionalized and design doomed to a tertiary support role. Only if design is recognized as a cognitive process will it have a chance to escape the straitjacket of the turn-of-the-century ideologies and take its place in the professional arena.

The acceptance of a new definition, namely of design as a cognitive function rather than a hand skill, is the first step in the evolution of design practice. In shifting position and heading into the only direction in which the field can reach its goals of professionalism, the next step is to embed the principles of cognitive design in the education of the new generation of practitioners. Through major curricular reforms, graphic designers and typographers must understand the full complexities of language, perception and human and social behaviour, not just in cursory ways via survey or introductory courses. They must be as comfortable with social and behavioural science as with literature and philosophy. Graduate students must want to understand the complexities of human communication and be able to employ research skills to build the information platform from which the profession can rise. The speed and success with which the profession is going to emerge is inextricably linked to the strength, depth and integrity of the research and the breadth of the accumulating database.

In 1978, at the ICOGRADA conference in Chicago, several of the planners, among them Jay Doblin, made exerted efforts to persuade design teachers and practitioners to recognize that the traditional design process spawned by French and Bauhaus design axioms of self-expression and concentration on form rather than contents and context had not protected the users from expensive failure of design. Still today, design education has made little effort to expand its responsibilities and vision of graphic design beyond the traditions.

Part of the underlying problem is that neither the Bauhaus nor other influential design schools have instilled in the design constituency a disciplined process of research, with the ethical understanding of all necessary skills, including ownership, authorship, verifiability and assurance of fidelity of information. What most designers understand as research is information gathering, sometimes information synthesis and analysis, but rarely as the testing of conceptual models, or the testing and application of data from findings in sociology or psychology. To many, the malfunctioning of design solutions has more to do with the choice of visual packaging than the design's misfit with complex social, economical, political and psychological contexts.

Some schools are responding to the needs for research, but the task for expansion of the curricular base is left to traditionally educated studio teachers who were all raised with the studio emphasis. The problems of the user environment cannot be solved merely by personal experience, or general knowledge, or market research. Although useful, market research and comparative product analysis are only a fraction of the total evaluation or planning process. Therefore, designers must learn to distinguish between claims based on conjecture and work based on empirical results of tests with users.

Today, in addition to the valuable visualization skills and concern for aesthetics introduced by the Bauhaus, the designers' education must include basic grounding in all aspects of research, the understanding of human acquisition of knowledge, retention and comprehension, as well as perception and language. The application of concepts of logic to establish project goals, criteria for evaluation and solving of problems must already be part of undergraduate education. Philosophy, logic and ethics courses will also help in developing the field's professional etiquette and ethics. Even at the beginning levels, designers must be taught to incorporate the understanding of human factors and user feedback into the design process. If the undergraduate education has certain cognitive components built into its programmes, then graduate education can expand and focus on original research, building a reservoir of knowledge. Only in this way can the design field expect to emerge as a profession.

The general ethical dysfunction of design education seems to be continued by the American design schools altogether. Its moral and political apathy is reflected in curricula that do not have room for discussions on values or value discrimination, ethics and morals. This is exacerbated by a design practice in which value is declared by the corporate client or marketing, rarely addressing ideals, but more often the lowest and most common denominators. In their jockeying for status and position, designers have aligned themselves with corporate management and are mimicking its language of efficiency and usefulness. Meanwhile, it is clear that their own information base, from which most of their decisions are made, is too narrow to provide leadership and guidance. Those qualities are now provided from outside the design field.

This might also be the reason for the distinct lack of a collective professional voice. The organizations that practitioners and students join are not ready to deal with ethical and moral issues such as racism, the homeless, the socially disadvantaged, or even visual pollution. Their inability to envision a better world, in quality or function, restricts them from materializing plans and tools for a more wholesome future. The intellectual apathy hinders the practice from providing the leadership for the evolvement of culture and the development of a philosophical base on which to build a socially, economically and environmentally sound existence.

Even during the politically most active time, the 1960s, when the USA's Vietnam, social and urban policies were challenged, its design schools did not provide shining examples of alternative ways to activate for peace or social justice. There were some that energized students to action, pressing volatile messages into clean corporate iconography. But that brief moment passed, and it was back to business as usual.

Absorbing the ideology of the Bauhaus as an idealized, apolitical entity into the fibre of both education and practice has given rise to irrational positions of neutrality in face of rather ethically complex issues. The notion that designers are neutral facilitators of communication is as faulty as that of information neutrality. Therefore, before the design process starts, the designers' ethical systems must determine if it is at all reasonable to support a project, not merely for the sake of financial survival, but because of its impact on society and culture.

The lack of research discipline, and its requirements for fidelity and integrity, have particularly affected design history and most publications on design. American educators and practitioners have noticed that vigorous public relations propelled the Bauhaus into the forefront of the field. Yet they have not similarly noted the absence of scholarship. Practitioners and educators have learned from the Bauhaus that appearing in print, in exhibition catalogues, books or articles, legitimizes their philosophical positions. In the competitive nature of the design practice, it is all too often forgotten that all professional endorsements are political, all organizations and juries are biased, and that there is no simple answer, truth or ultimate aesthetic. At the Academy, students were taught to pay attention to the sources of knowledge. They learned to distinguish among fact, conjecture and purely personal opinion. Since the Bauhaus and before, design education has conveniently forgotten authors and origins of ideas, and this lack of scholarly discipline will haunt the design field for years to come. Without a research base, it is impossible to proclaim the efficiency, communication fidelity or user satisfaction brought about by so-called 'good design'. The slogans of 'good design sells' and 'form follows function' are hollow without substantiation. In the family of professions, each assertion finds resonance only if it is followed by verification.

The lack of design research deprives the field of the advantage of being ahead of other disciplines in discovery and foreshadowing of events. Without the inner strength to lead others or the ability to choose and guide themselves, schools depend on the advice of successful practitioners and alumni. Although it is true that practitioners accrue experiential knowledge, their design methodology, however, is linked to the specific success of their own approach and interests as well as business opportunities. Frequently, their vision of design education reflects the specific narrowness of their own talents. If they succeed on the basis of intuitive design, without benefits of cognitive analysis, their vision of a good curriculum does not include any additional components to those that provided them with success. In their minds, if any change is required at all, it is to expand the singularly beneficial components. In the need to survive financially and to place graduates in the field, most schools are obliged to heed the advice from their direct constituents. Many schools go even further and appoint their successful alumni to the faculty. This incipient process, in which schools place graduates with practitioners, who then return as teachers, who base their teaching on the narrow models of their own experiences, who then participate in rewriting the curriculum to fit vocational needs, is a downward spiral in which intellectual or cognitive growth cannot be developed.

Design schools have to take a serious look at themselves and decide if they are truly capable of supporting a design profession. They will have to select between either promulgating the safe, the redundant and the traditional, or escaping the limitations of long-standing traditions and redefining design, adding research and cognitive components to their course menus. The hope, of course, is that their choice will come down on the side of an information-rich, contextual, culturally and socially responsible, communications-oriented design process. If not, then there is the real danger that the field will step aside and leave the shaping of the future to other disciplines.

As teachers and practitioners, will we be satisfied to just repackage the Bauhaus for the future by sprucing it up and putting lipstick on its face, or are we willing to contribute the energy to shape the contents and the context of design education, and in doing so evolve a profession? The present trend to just link students up with business programmes to introduce them to marketing and commerce practices is a welcome expansion but does not go far enough. In reality, design must be connected with all aspects of communication and all the complex specific and general issues that arise from the various media and the contexts where they operate.

Bibliography

Note: Where the source cited is not in English, the quotation has been translated by the author.

The year of publication and page numbers entered in the body of the text correspond to the edition available at the time of writing this book, and not necessarily to the original edition. Where the information was available, reference to the year of original publication appears in this bibliography. For multi-authored materials, the reference in the text lists only the first author.

ADVERTISING FEDERATION OF AUSTRALIA (1990) *Effective Advertising*, Casebook of the AFA Advertising Effectiveness Awards, Sydney, Australia.

ALBERTA TRANSPORTATION AND UTILITIES (1990) *Alberta Traffic Collision Statistics*, Edmonton.

ALEXANDER, CHRISTOPHER (1979), *Notes on the Synthesis of Form,* 1st Edn 1964, Cambridge, MA: Harvard University Press. Copyright 1964 by the President and Fellows of Harvard College. Reprinted by permission of Harvard University Press.

AXELROD, MYRIL (1975) Ten essentials for good qualitative research, in James B. Higginbotham, and Keith C. Cox (1979), *Focus Group Interviews: A Reader,* pp. 55–9, Chicago: American Marketing Association.

BACHELARD, GASTON (1948) *La Terre et les rêveries de la volonté*, Paris: Corti.

BARTHES, ROLAND (1985) *The Responsibility of Forms*, New York: Hill & Wang.

BATESON, GREGORY (1972) *Steps to an Ecology of Mind*, New York and Toronto: Ballantine Books.

BEIRNESS, D.J. AND SIMPSON, H.M. (1987) Alcohol use and lifestyle factors as correlates of road crash involvement amongst youth, in T. Benjamin (ed.), *Young Drivers Impaired by Alcohol and Other Drugs*, pp. 141–8, London Royal Society of Medicine Services: London.

BELLENGER, D. N., BERNHARDT, K.L. AND GOLDSTUCKER, J.L. (1976) Qualitative research techniques: focus group interviews, in James B. Higginbotham and Keith C. Cox (1979), *Focus Group Interviews: A Reader*, pp. 13–33, Chicago: American Marketing Association.

BERNARD, PIERRE (1990) Nobody talks about politics: what makes graphic design a social act?, in Jorge Frascara (ed.) *Graphic Design, World Views*, pp. 178–82, Tokyo: Kodansha.

BOCCHI, G. AND CERUTI, M. (eds) (1992) *La Sfida della Complessità,* Milano: Feltrinelli.

BONSIEPE, GUI (1967) Arabesken der Rationalität/Arabesques of Rationality', *ulm* **19/20**, FfG Ulm, 9–23. Reprinted (1989) as Arabesques of rationality: notes on the methodology of design, in *Readings from Ulm*, pp 145–59, Bombay: Industrial Design Centre.
BRINBERG, DAVID AND MCGRATH, JOSEPH E. (1985) *Validity and Reliability in the Research Process,* Beverly Hills, CA: Sage Publications.
BRONOWSKY, JAKOB (1973) *The Ascent of Man*, London: BBC.
BURKE, KENNETH (1957) *The Philosophy of Literary Form*, New York: Vintage Books.
CHAKRAPANI, CHUCK AND DEAL, KENNETH (1992) *Marketing Research: Methods and Canadian Practice*, Scarborough, ON: Prentice Hall Canada Inc.
CONVERSE, JEAN M. AND PRESSER, STANLEY (1986) *Survey Questions: Handcrafting the Standardized Questionnaire*, Beverly Hills, CA: Sage Publications.
CROSS, NIGEL AND ROY, ROBIN (1975) *Design Methods Manual*, Milton Keynes: Open University.
CUTTS, MARTIN AND MAHER, CHRISTINE (1981) Simplifying DHSS forms and letters, *Information Design Journal* **2**(1), 28–32.
DEJONG, W. AND WINSTEN, J.A. (1989) *Recommendations for Future Mass Media Campaigns to Prevent Pre-teen and Adolescent Substance Abuse*, Boston: Center for Health Communication, Harvard School of Health.
EAGLESON, ROBERT (1983) *The Reform of Documents within Government Departments in the United Kingdom,* Canberra: Information Coordination Branch.
EDMONSTON, PHIL (1990) *Lemon-Aid Used Car Guide*, Toronto: Stoddart.
EHSES, HANNO (1986) *Design Papers 4,* Halifax: Nova Scotia College of Art and Design.
EHSES, HANNO (1988) *Design Papers 5*, Halifax: Nova Scotia College of Art and Design.
EPSTEIN, JOYCE (1981) Informing the elderly, *Information Design Journal* **2** (3/4), 215–35.
FINN, ADAM (1992) Driving and driver behaviour: report on focus groups with young male adults, Appendix 2, in Jorge Frascara, Adam Finn, Henri L. Janzen, John G. Paterson and Zoe Strickler-Wilson, *Traffic Safety in Alberta/Casualty Collision and the 18–24 Year Old Male Driver: Criteria for a Targeted Communication Campaign*, Edmonton: Alberta Motor Association/Alberta Solicitor General.
FIRTH, DIANE (1981) An investigation of the success of redesigned supplementary benefit documents, *Information Design Journal,* **2** (1), 33–43.
FISHER, PHIL AND SLESS, DAVID (1990) Information design methods and productivity in the insurance industry, *Information Design Journal* **6** (2), 103–29.
FRASCARA, JORGE, FINN, ADAM, JANZEN, HENRI L., PATERSON, JOHN G. AND STRICKLER-WILSON, ZOE, (1992) *Traffic Safety in Alberta/Casualty Collision and the 18–24 Year Old Male Driver: Criteria for a Targeted Communication Campaign*, Edmonton: Alberta Motor Association/Alberta Solicitor General.
FRISCH, O. R. (1979) *What Little I Remember*, New York: Cambridge University Press. Reproduced with permission.
GOLDMAN, ALFRED (1962) The group depth interview, in James B. Higginbotham, and Keith C. Cox, (1979) *Focus Group Interviews: A Reader,* pp 43–50, Chicago: American Marketing Association.
GROPIUS, WALTER (1962) *Scope of Total Architecture*, New York: Collier Books. Quotations by permission from Harper Collins Publishers, *The Scope of Total Architecture, World Perspectives*, Vol. 3, by Walter Gropius. Copyright 1943, 1949, 1952, 1954, 1955 by Walter Gropius.
HARPER, GREG (Grey Advertising) AND L'HUILLIER, LEON (Transport Accident Commission) (1990) Road safety: the Transport Accident Commission campaign, in *Advertising Effectiveness Awards Casebook 1*, North Sydney: Advertising Federation of Australia.

HIGGINBOTHAM, JAMES B., AND COX, KEITH C. (1979) *Focus Group Interviews: A Reader*, Chicago: American Marketing Association.

HOFSTÄDTER, DOUGLAS (1980) *Gödel, Escher, Bach: An Eternal Golden Braid*, New York: Vintage Books.

KIRK, JEROME AND MILLER, MARC L. (1956) *Reliability and Validity in Qualitative Research,* Sage University Paper Series on Qualitative Research Methods, Vol. 1, Beverly Hills, CA: Sage Publications.

LE MOIGNE, JEAN-LOUIS (1992) 'Progettazione della Complessità e Complessità della Progettazione', in Bocchi, G. and Ceruti, M. (eds), *La Sfida della Complessità*, Milano: Feltrinelli.

LEWIS, DAVID (1986) The cost of errors in forms, *Information Design Journal* **4** (3), 221–3.

LICHTENBERG, G.C. (1844–53) *Vermischte Schriften*, Goettingen.

MALDONADO, TOMÁS (1959) Communication and semiotics, *Ulm 5*. Reprinted (1989) in *Readings from Ulm*, Bombay: Industrial Design Centre.

MALDONADO, TOMÁS (1974) *Avanguardia e Razionalità*, Torino: Einaudi.

MALDONADO, TOMÁS (1989) The emergent world: a challenge to architecture and industrial design training, in *Readings from Ulm*, Bombay: Industrial Design Centre. Originally published in Spanish (1966) as Hacia una proyectación del ambiente, *Summa 6/ 7*. Published in Italian (1974) as Verso una progettazione ambientale, in *Avanguardia e Razionalità*, Torino: Einaudi.

MALDONADO, TOMÁS (1992), *La Speranza Progettuale*, 1st Edn 1970, Torino: Einaudi. Published in English (1972) as *Design, Nature and Revolution,* New York: Harper & Row.

MALDONADO, TOMÁS (1993) *Reale e Virtuale*, Milano: Feltrinelli.

MARGOLIS, HOWARD (1987) *Patterns, Thinking and Cognition*, Chicago: University of Chicago Press.

MCCRACKEN, GRANT (1988) *The Long Interview,* Sage University Paper Series on Qualitative Research Methods, Vol. 13, Beverly Hills, CA: Sage Publications.

MCGUIRE, WILLIAM J. (1985) Attitudes and attitude change, in G. Lindzey and E. Aronson (eds) *The Handbook of Social Psychology,* Vol. 2, 3rd Edn, pp. 233–346, New York: Random House.

MICHMAN, RONALD D. (1991) *Lifestyle Market Segmentation*, New York: Praeger Publishers.

MILES, MATHEW B. (1979) Qualitative data as an attractive nuisance, in J. Van Maanen (ed.) *Qualitative Methodology*, pp. 117–32, Beverly Hills, CA: Sage Publications.

MILES, MATHEW B. AND HUBERMAN, A. MICHAEL (1984), *Qualitative Data Analysis: A Sourcebook of New Methods*, Beverly Hills, CA: Sage Publications.

MOLES, ABRAHAM (1964) Le Contenu d'une méthodologie appliqué, in *Méthodologie – vers une science de l'action*, Paris: Gauthier.

MORIN, EDGAR (1992) *Il Metodo*, 1st Edn 1977, Milano: Feltrinelli.

MORGAN, DAVID L. (1988) *Focus Groups as Qualitative Research,* Sage University Paper Series on Qualitative Research Methods, Vol. 16, Beverly Hills, CA: Sage Publications.

NÄÄTÄNEN, RISTO AND SUMMALA, HEIKKI (1976) *Road-User Behaviour and Traffic Accidents*, Amsterdam: North Holland and New York: American Elsevier. Reproduced with permission.

NATIONAL COMMITTEE FOR INJURY PREVENTION AND CONTROL (1989) *Injury Prevention: Meeting the Challenge*, New York: *American Journal of Preventive Medicine* and Oxford University Press.

NATIONAL SAFETY COUNCIL (1992) *Accident Facts,* Itasca, IL.

OGILVY, DAVID (1983) *Ogilvy on Advertising*, New York: Crown.

PLATEK, R., PIERRE-PIERRE, K. AND STEVENS, P. (1985) *Development and Design of Survey Questionnaires*, Ottawa: Statistics Canada, Census and Household Survey Methods Division.

RAPOPORT, ANATOL (1962) The use and misuse of games theory, *Scientific American*, December, 108–18.

RICE, D.P., MACKENZIE, E.J. AND ASSOC. (1989) *Cost of Injury in the United States: A Report to Congress,* San Francisco: Institute of Health & Aging, University of California, and Injury and Prevention Center, The Johns Hopkins University.

ROAD SAFETY AND MOTOR VEHICLE REGULATION DIRECTORATE (1987) *Smashed*, Ottawa: Transport Canada.

ROTHE, J. PETER (ed.) (1987) *Rethinking Young Drivers*, Vancouver: Insurance Corporation of British Columbia.

SANDAY, PEGGY REEVES (1983) The ethnographic paradigm(s), in J. Van Maanen (ed.) *Qualitative Methodology*, pp. 19–25, Beverly Hills, CA: Sage Publication.

SAUSSURE, FERDINAND DE (1915) *Cours de linguistique générale*, Geneva. Published in English (1959) as *Course in General Linguistics,* New York: Philosophical Library, and (1966) New York: McGraw-Hill.

SELVINI PALAZZOLI, M. (1989) *Sul Fronte dell'Organizzazione*, Milano: Feltrinelli.

SLESS, DAVID (1985) Informing information designers, *Icographic* **2** (6).

STATISTICS CANADA (1990) *Catalogue 13-207, Income Distributions by Size in Canada,* Ottawa.

STENGERS, ISABELLE (1992) Perchè non può esserci un paradigma della complessità, in G. Bocchi, and M. Ceruti (eds) *La Sfida della Complessità*, pp. 61–83, Milano: Feltrinelli.

SUDMAN, SEYMOUR AND BRADBUM, NORMAN M. (1982) *Asking Questions: A Practical Guide to Questionnaire Design*, San Francisco, CA: Jossey-Bass Publishers.

SZENT-GYORGI, ALBERT (1972-80) *System and Structure,* London and New York: Tavistock Publications.

University of Reading (1975) *Graphic Communication through Isotype,* Reading.

VAN MAANEN, JOHN (ed.) (1983) *Qualitative Methodology,* Beverly Hills, CA: Sage Publications.

VERCELLONI, VIRGILIO (1987), *L'Avventura del Design: Gavina*, Milano: Jaca Book.

WALLER, ROBERT (ed.) (1981) The design of forms and official information, *Information Design Journal* **2** (3 and 4).

WALLER, ROBERT (1985) Designing a government form: a case study, *Information Design Journal* **4** (1), 36–57.

WILDEN, ANTHONY (1980) *System and Structure*, 1st Edn 1972, London and New York: Tavistock Publications.

WILDEN, ANTHONY (1987) *The Rules Are No Game*, London and New York: Routledge & Kegan Paul.

WITTGENSTEIN, LUDWIG (1961) *Tractatus Philosophicus*, London: Routledge & Kegan Paul. First published in German (1921) in *Annalen der Naturphilosophie*.

WÖLFFLIN, HEINRICH (1915) *Kunstgeschichtliche Grundbegriffe*, Munich: F. Bruckmann A.G. Published in English (1932) as *Principles of Art History*, London: G. Bell & Sons Ltd, and (1950) by Dover Publications.

WORRINGER, WILHELM (1911) *Abstraktion und Einfühlung*, Munich: R. Piper. Published in English (1953) as *Abstraction and Empathy,* New York: International Universities Press.

WORRINGER, WILHELM (1920) *Formprobleme der Gotik,* Munich: R. Piper. Published in English (1927) as *Form in Gothic,* London: G. P. Putnam's Sons, Ltd.

WURMAN, RICHARD SAUL (1976) *What-if, Could-Be: An Historical Fable of the Future*, author's edition, Newport, RI.

Index

accident
  meaning of term 64–5, 77
  see also injuries
accountability of designer 21–3
accuracy
  measurement 44–5
  transcript 57
action 120
advertising and posters 2, 25
  airlines 27
  alcohol 62, 80–3
  cars 62, 83–9
  driving with care 101–5, 112–15
  Poland 14, 25–6
  stereotypes 27–8
aesthetics 13–14
  of order 30–1
Africa 21, 22
age and Alberta traffic safety project 65–6
aircraft and airlines
  advertising 27
  safety 12–13, 16, 24
Albers, Josef 131
Alberta: driving and road safety 10, 16; *see also* targeting communications
alcohol
  advertising 62, 80–3
  and driving 90, 92, 107, 108
  excluded in Alberta project 67
alertness of drivers 75–6
Alexander, Christopher 27, 33, 40
Amazon 29
'anchoring' text 41–2
anthropology 7
appropriate method 36
architecture 26–7, 131
Argentina: bureaucracy 13, 20, 31
arts, fine 13, 19
attitudes and behaviours
  affecting, selling products vs 4–5
  behavioural change as objective *see* driving and road safety
  behaviouristic ethnography 47
  formation 47
  revealing *see* focus groups; targeting communications
attractiveness of visual communication design 12, 13–14
audience 4, 8–11
  segmentation 8–9, 109 *see also* targeting communications
Australia
  Capita Insurance Company forms 116–17
  traffic safety *see* Transport Accident Commission
authority-based communication 17–18, 28
avant-garde 13
Axelrod, Myril 48, 49–50

Bachelard, Gaston 25
bad driver concept 75–6
Banks, Colin 117
Baroque 30, 131
Barthes, Roland 41, 91
Bateson, Gregory 25, 42
Bauhaus 8, 14, 129–35
Bayer, Herbert 14
Beck, Ulrich 126
behaviour *see* attitudes and behaviours
Beirness, D.J. 67
believability of visual communication design 12, 15–17
Bellenger, D.N. 47
Benedict, Ruth 46
Beresh, Bryce 101
Bernard, Pierre 18, 25
birth control 15–16, 22
BMW 72

Boas, Franz 46
Bohr, Niels 11
Bonsiepe, Gui 40
Bourdieu, Pierre 128
Bradbum, Norman M. 52
brain and mind 35
brand loyalty 109
Braun 14
Breuer, Marcel 26
Brinberg, David 44, 46
Britain
  British Telecom telephone book 117–18
  causes of death 64
  DHSS forms 19, 112–16
Bronowsky, Jakob 30
budgets *see* expenditure/costs
Buick 84, 85
Burke, Kenneth 15
Byfield, Stuart 112, 113, 114

Cadillac 67–8, 84, 86
Camarao 67
Canada
  advertising 28
  hockey helmets 111
  illiteracy 23
  injuries and deaths 20, 21, 64
  road safety *see* Alberta
Capita Insurance Company forms (Australia) 116–17
Carlson, Rey 113, 114
Carnap, R. 38
cars
  advertising 62, 83–9
  sizes and types 67–8
  see also driving and road safety
case histories 24, 107–18
  Australia *see* Capita Insurance; Transport Accident Commission
  British Telecom telephone book 117–18
  DHSS forms in Britain 19, 112–16
  see also targeting communications
casualties *see* injuries
categories defined 43
causes of accidents, focus group's opinion of 76–9
Chakrapani, Chuck 9
change, concept of 119, 120
city 29–30, 124–5
closed-ended questions 54–6
clothing 26
codes of conduct 18
codes (linguistics) 28, 30
cognition 37
Communications Research Institute (Australia) 116
comparison of research findings 46
complexity 15, 34
composition 42
computers 7–8, 121–2, 123, 125
control and driving 99
Converse, Jean M. 52, 53, 56
convincingness *see* believability
Coopers & Lybrand Associates 116
Corvette 67, 68
costs *see* expenditure/costs
courtesy of drivers 75, 76
crashes, road *see* driving and road safety; driving *under* injuries; targeting communications
Creapole ESDI 122
Cross, Nigel 3, 35
Cubana Airlines 12–13
cultural responsibility of designer 24–8
Cutts, Martin 115

daily life 26–7, 35
dated advertisements recognized 40
Deal, Kenneth 9
deaths *see* injuries and deaths
definition of problem 39
DeJong, W. 63
Demchuk, Jerry 102
demographic of segmentation of audience 8
Department of Health and Social Security forms 19, 112–16
depth interview 50
deschooling and design education 126–9
design *see* user-centred graphic design
designer 11–28, 119–35
  accountability 21–3
  cultural responsibility 24–8
  deschooling and design education 126–9
  ethical responsibility 17–18
  multidisciplinary coordination 5–6
  political capacity 6
  practice and education beyond Bauhaus 129–35
  professional responsibility 12–17
  social responsibility 19–24
  transformation of design 119–26
detectability of visual communication design 12
DHSS (Department of Health and Social Security) forms 19, 112–16
dimension of Alberta traffic safety problem 63–4
disability 19–20
disciplines and interdiscipline 5–8, 126, 134
discriminability of visual communication design 12
disease *see* safety and health
Doblin, Jay 133
Document Design Unit (UK) 116
Doolan, Jamie 112
doubt 126
Dow, Rob 112, 115
driving and road safety 10, 16

high-collision intersection 33–4
injuries and deaths 20–1, 22, 63, 67, 107
see also cars; targeting communications

Eagleson, Robert 116
Earth Summit 125
ecological cycle of product 122
Edmonston, Phil 67
education
design 6–8, 126–35
driver training 79, 98
materials 15–16, 22
effective communication 3–4
efficiency valued 31
Ehses, Hanno 38
Elturan, Nurhayat 102
emotional
appeal and persuasion 96–7
components of driving 99
enforcement of laws 16, 67
environment 122
epidemiological approach 10–11, 63
Epstein, Joyce 19, 114
error in interviews and surveys 52–3
Eskimos 29, 46
ethical responsibility, designer 17–18
ethnographic research 46–7, 50
European Union
causes of death 64
telework 123
see also Britain; Germany
evaluation of performance 5, 22, 98
expenditure/costs
on cars 74–5
on design 19
on health care 20–1, 63
on prevention 63
on redesigning forms 116
on traffic crashes 63
on traffic safety projects 68–9, 109, 111

failures in research as information sources 46
fatalism of drivers 77, 80
Finn, Adam 62, 71–80, 88–89
Firth, Diane 115
Fisher, Phil 115, 116
fitness of method to problem 35–7
flexibility 35
focus groups 47–50
and Alberta traffic safety project 62, 69–71
graphic communications, reactions to 80–93
questions and responses 71–80
selection of subjects 69
sessions 71–93
setting up 69–71
analysis and reporting 50
sessions 49, 71–93
setting 48–9

form-giver *see* designer
forms improved 19, 112–17
France
causes of death 64
computers 122
Francescutti, Dr Louis 10
Frascara, Jorge vi, ix–x
Frauenhofer Institute 124
Frisch, O.R. 11
function and aesthetics 14
furniture 26

Gallup Poll 43
game theory 34
Gavina, Dino 27
gender and Alberta traffic safety project 65–6, 99
generalizability of social research 44
geographic segmentation of audience 8
Germany
causes of death 64
social inequality 125
telework 123
'Terravision' project 121
see also Bauhaus
gestalt theory 15
globalization 123–5
Goldman, Alfred 47, 48, 49
good design concept 40
good driver concept 75
Gothic 30
graphic design *see* user-centred graphic design
Grapus 18
Grey Advertising 16, 107–8
Gropius, Walter 8, 12, 21, 131, 132
groups 21
see also focus groups
Guyer, Ben 103

Harper, Greg 112, 113, 114, 115
Harvard University 132–3
health *see* safety and health
hierarchizing problem 40
high-risk drivers *see* targeting communications
Hitler, Adolf 131
hockey helmets 111
Hofstädter, Douglas 29
holistic ethnography 47
homogeneity of focus groups 48
Huberman, A. Michael 56, 58
humour in advertising, inappropriate 89, 91
Humphrey, Mat 115
hypotheses 5, 45

ICOGRADA conference 133
iconography 96
identification
of problem 20–1, 39
Alberta traffic safety project 61–2

with victims 109
images
and texts 41–2
and values 28
India: networking 123–4
indices to measure success of strategy 10
information
access as right 23
generating qualitative *see* focus groups
and persuasion 37–8
from research failures 46
signs 37–8
interpreting 29–30
as symbols 38
without 25
technology *see* computers; networking
Information Design Journal 15
injuries and deaths
and age 65
and disease 63, 64
driving 20–1, 22, 63, 67, 107
Institute for Economics and Society 123
instructional materials 15–16, 22
International Parenthood Federation (Africa) 22
interviews 50–2
group *see* focus groups
intuition 11
invisibility of design 27
Isotype 21
issues, sorting 42–3
Italy: parking signs 16

Janzen, Henri L. 62, 93
Japan: causes of death 64
Jeep 84, 89
jibaros 29
John of the Cross, St 36
Johnson, Dr Michael 109
journalism *see* mass media

Kenya: Masai 30
Kepes 132
Kinneir, Jock 37
Kirk, Jerome 45, 46
Knoll International 14, 17

labelling 13
language
mastery of 34
natural 38, 47
plain and simple 53
improved in forms 19, 112–17
in traffic campaign 97–8
law *see* legislation and law
Le Moigne, Jean-Louis 28, 34
learning and understanding 29
Leavitt, J.H. 38
legislation and law
labelling 13
lawful behaviour assumed 34
seat-belts 16, 67
see also offending drivers
Lewis, David 115, 116
L'Huillier, Leon 108
Lichtenberg, G.C. 36
life
improving 20, 23–4
making possible 20–3
see also safety
lifework concept 119–20
light 42
linguistics 28, 30, 37
Lissitsky, El 13
literacy 23
long interview 50–2

McCracken, Grant 43–4, 47, 50, 57
McGrath, Joseph E. 44, 46
McGuire, William J. 99
Maher, Christine 115
Maldonado, Tomas 6, 24, 25–6, 28, 38
male drivers, young *see* targeting communications
Malinowski, Bronislaw 46
mapping terrain 3–32
audience 4, 8–11
discipline and interdiscipline 5–8
meaning, order and freedom 28–30
working profile 3–5
see also designer
Margolis, Howard 35
market economy 19
marketing 7, 109
research 12
Masai 30
mass media 9, 25, 37–8
campaign, planning *see* targeting communications
choosing right 3
impact of Australian driving project on 109
see also advertising and posters
Mays, W. 38
Mead, Margaret 43, 46
meaning
and order 28–30
and perception 15, 28–9
measurement
accuracy 44–5
of effects 40
measurable audience 9, 10
quantitative research 33–7, 43, 44
of success 10
media *see* mass media
Mercedes-Benz 124
Mercury Lynx 88
metaphors 57
methods, design 33–60
fitness to problem 35–7

markers 39–40
quantifiable, human dimension and 33–7
requirements, sorting 42–3
semiotics, insufficiency of 37–9
visualization of strategies 40–2
see also validity in data collection
Meurer, Bernd x
on transformation of design 119–26
Meyer, Hannes 131
Meyer-Eppler, W. 38
Michman, Ronald D. 9
Mies van der Rohe, Ludwig 31, 131
Miles, John 117
Miles, Mathew B. 56–7, 58
Miller, Marc L. 45, 46
Moles, Abraham 35
Morgan, David L. 48, 49
Morin, Edgar 35, 36
multidisciplinary coordination 5–6

Näätänen, Risto 64, 76, 79, 94
Nachado, Antonio 35
Nagy 132
names, magical power of 37
National Association of German Industry 123
National Committee for Injury Prevention (US) 22, 63
National Safety Council (US) 63, 65
naturalism (realism) 44, 109
Negroponte, Nicholas 121–2
Netherlands: causes of death 64
networking 122, 123–4
Neurath, Otto 23
neutral questions 53
New Bauhaus 132
newspapers *see* mass media
Nissan 84, 85
North America *see* Canada; United States

objectives
of traffic safety project 95
of visual communication 3–4
objectivity problems 45–6
see also validity
objects, culture of 26–8, 120
obscure communication as abuse 15
offending drivers
focus group in Alberta traffic safety project 69–70
rehabilitation attempts 10
Ogilvy, David 96
Oh, Sungmi 103
open-ended questions 54–6, 62
order
aesthetics of 30–1
and meaning 28–30
Other as subject 17
Otis 124

Pakistan: birth control 15–16
Palazzoli, Selvini 36
participant observation 46–7
partner, public as 18
Paterson, John G. 62, 93
path, method as 35
Paulham, F. 38
peer groups and young drivers 73
Pentagram 16
perception
basic principles 4
and meaning 15, 28–9
of public benefit 5
of risk 76–7
and survival 15
personality
profile of high-risk driver 93–4
types 9
persuasion and information 37–8
Platek, R. 52, 54, 56
Plymouth 88
point of view in image 42
Poland: posters 14, 25–6
politics/political
capacity of designer 6
speeches 25
Pontiac 87
posters *see* advertising and posters
power and driving 99
precision 34–5, 44
Presser, Stanley 52, 53, 56
problem-solver and identifier, designer as 20
professional responsibility of designer 12–17
'projective questions' 49
prototype messages 39
psychology 7, 9
change and high-risk driving 99
public benefit, perception of 5

qualitative research 43–4
data analysis 56–8
see also focus groups; interviews
quantitative research 43, 44
quantifiable methods 33–7
questions
in driving focus groups 62, 71–80
see also interviews; survey questionnaire

rainforest 21, 29
Rams, Dieter 14
Rapoport, Anatol 34
rational analysis 11
reachable audience 9
reactivity 9, 10, 19
realism (naturalism) 44, 109
rehabilitation of offending drivers attempted 10
'relaying' text 41, 42
relevance 25

reliability of research 45–6
Renaissance 11, 30
repeatability of observation 36
research 12, 44–9, 50, 116
  lack of 134–5
  Research Institute for Consumer Affairs (UK) 19, 114, 115
  see also case histories; methods, design; qualitative research
rhetoric and semiotics 38
Rice, D.P. 63
Rio Earth Summit 125
road safety *see* driving and road safety
Rothe, J. Peter 71
Roy, Robin 3, 35

safety and health 20–1, 63
  airlines 12–13, 16, 24
  diseases, deaths from 63, 64
  posters 88–93
  see also driving and road safety
Salford Form Market 115
Samoa 43
Sanday, Peggy Reeves 47
Sassen, Saskia 124
Saussure, Ferdinand de 37
Schmidt, Peter 112, 114
seat-belts 16, 67
segmentation, audience 8–9, 109
self-esteem of drivers 72–3, 93, 96
'self-reported' data 48
selling products vs affecting attitudes 4–5
semiotics 41–2, 91
  ethnography 47
  insufficiency of 37–9
service-oriented principles 122–3
Shannon, Claude Elwood 17
Shapiro, Stacey 104
significant utterances 57
signs *see under* information
simplicity
  of design *see* Bauhaus
  of language *see* plain *under* language
Simpson, H.M. 67
Sless, David 115, 116, 117
social handicap 23
social inequality 125
social perspective of researcher 45, 48–9
social responsibility of designer 19–24
social space 125
socioeconomic segmentation of audience 8
sociology 7
Solidarity 25
Sony Walkman 12
source of message and believability 16
Soviet Union
  advertising 28
  aircraft from 12–13
  government buildings 26
specificity of design problem 38–9
speed 124
  and traffic safety projects 71, 72, 78, 99–100, 107, 108
Steadman & Ryan 110
stereotypes 27–8, 74
strategies
  appropriate 36
  road safety 95–8
  success, measuring 10
  visualization of 40–2
Strickler-Wilson, Zoe x, 62
  on validity in data collection 43–59
style 42
substantial audience 9, 10
success of campaigns 110
Sudman, Seymour 52
Summala, Heikki 64, 76, 79, 94
survey questionnaire 52–6
  asking questions 53
  flow and layout 56
  long interviews 51–2
  types of question 54–6
survival 15
sustainable development, design as 125
sustained continuation of scheme 40, 98–9
Sweeney (Brian) & Associates 108
symbols/symbolism
  cars and driving 93–4, 97
  signs as 38
Szent-Gyorgi, Albert 15

targeting communications (Alberta traffic safety project) 40–1, 61–106
  campaign concept 99–100
  development steps 39–40, 61
  future action 40, 98–9
  objectives 95
  problem identification 61–2
  recommendations for communication campaign strategy 95–8, 100
  target group
    defining 64–7
    narrowing down 93–4
    profile of 68–71
    see also focus groups *under* Alberta
  visualizing ideas 100–5
Tejada, Barbara 105
television *see* mass media
telework and telecommunications 120–1, 123–5
'Terravision' project 121
testing 39
texts and images 41–2
themes of traffic campaign 95–6
things *see* objects
Thresher, Paul 109
time 30–1, 119–20, 121, 124
tone of traffic campaign 97

Toorn, Jan van x–xi
on design education 126–9
Touraine, Alain 125
toxicity and labelling 13
Toyota 68, 84, 86
traffic *see* driving and road safety
transformation of design 119–26
transmitter and receiver 17
transport 120, 124;
see also aircraft; cars; driving and road safety; targeting communications
Transport Accident Commission campaign (Australia) 16, 19, 64, 90, 107–12
attitudes and behaviours 111
conception 108–9
cost-effectiveness 110
results 110
strategy 110–11
truth in interviews and surveys 52
tuaregs 29

Ulm school 40, 131
understanding
and learning 29
visual communication design 12, 15
United Kingdom *see* Britain
United States
advertising 28
design education 129–35
telework 123, 124
traffic injuries and deaths 21, 22, 63, 67, 102
Vietnam war 17, 21, 134
usability 34–5
user-centred graphic design *see* case histories; designer; mapping terrain; methods; targeting communications

validity in data collection 43–59
interviews 50–2
methodologies in social sciences 43–4
participant observation 46–7
qualitative data analysis 56–8
see also focus groups; survey questionnaire
values and images 28
verbal articulation 11
Vercelloni, Virgilio 27
Vietnam war 17, 21, 134
Virilio, Paul 124
visualization of strategies 40–2
Volvo 68, 87

Waller, Robert 115
war 17, 21, 63, 134
Warnecke, Karen 109
Wilden, Anthony 2, 15
Winkler, Dietmar xi, 17
on design practice and education 129–35
Winsten, J.A. 63
Wittgenstein, Ludwig 28
Wölfflin, Heinrich 30
Worringer, Wilhelm 30
Wupperthal Institute 122
Wurman, Richard Saul 29

Yuen, Selene 104